Analysis of Engineering Cycles

WORKED PROBLEMS

Power, Refrigerating and Gas Liquefaction Plant

THERMODYNAMICS AND FLUID MECHANICS SERIES
General Editor: W. A. WOODS

Other Titles of Interest

BENSON
Advanced Engineering Thermodynamics, 2nd Edition

BENSON & WHITEHOUSE
Internal Combustion Engines

DANESHYAR
One-Dimensional Compressible Flow

DIXON
Fluid Mechanics, Thermodynamics of Turbomachinery, 3rd Edition

DIXON
Worked Examples in Turbomachinery (Fluid Mechanics and Thermodynamics)

DUNN & REAY
Heat Pipes, 3rd Edition

GOSTELOW
Cascade Aerodynamics

HAYWOOD
Analysis of Engineering Cycles, 3rd Edition

REAY & MACMICHAEL
Heat Pumps, 2nd Edition

Pergamon Related Journals (Free sample copy gladly sent on request)

Computers & Fluids

International Journal of Engineering Science

International Journal of Heat and Mass Transfer

International Journal of Multiphase Flow

Journal of Heat Recovery Systems

Analysis of Engineering Cycles

WORKED PROBLEMS

Power, Refrigerating and Gas Liquefaction Plant

By

R. W. HAYWOOD

Emeritus Reader in Engineering Thermodynamics
University of Cambridge

PERGAMON PRESS

OXFORD · NEW YORK · BEIJING · FRANKFURT
SÃO PAULO · SYDNEY · TOKYO · TORONTO

U.K.	Pergamon Press, Headington Hill Hall, Oxford OX3 0BW, England
U.S.A.	Pergamon Press, Maxwell House, Fairview Park, Elmsford, New York 10523, U.S.A.
PEOPLE'S REPUBLIC OF CHINA	Pergamon Press, Qianmen Hotel, Beijing, People's Republic of China
FEDERAL REPUBLIC OF GERMANY	Pergamon Press, Hammerweg 6, D-6242 Kronberg, Federal Republic of Germany
BRAZIL	Pergamon Editora, Rua Eça de Queiros, 346, CEP 04011, São Paulo, Brazil
AUSTRALIA	Pergamon Press Australia, P.O. Box 544, Potts Point, N.S.W. 2011, Australia
JAPAN	Pergamon Press, 8th Floor, Matsuoka Central Building, 1-7-1 Nishishinjuku, Shinjuku-ku, Tokyo 160, Japan
CANADA	Pergamon Press Canada, Suite 104, 150 Consumers Road, Willowdale, Ontario M2J 1P9, Canada

First edition 1986

Library of Congress Cataloging in Publication Data

Haywood, R.W. (Richard Wilson)
Analysis of engineering cycles, worked problems.
(Thermodynamics and fluid mechanics series)
Bibliography: p.
1. Power-plants—Problems, exercises, etc.
2. Refrigeration and refrigerating machinery—Problems, exercises, etc. 3. Gases—Liquefaction—Problems, exercises, etc.
I. Title. II. Series.
TJ164.H384 1986 621.4 86-9326

British Library Cataloguing in Publication Data

Haywood, R. W.
Analysis of engineering cycles, worked problems: power, refrigerating and gas liquefaction plant.—(Thermodynamics and fluid mechanics series)
1. Power-plants
I. Title II. Series
621.4 TJ164
ISBN 0-08-032572-6 Hard cover
ISBN 0-08-032571-8 Flexicover

Printed in Great Britain by A. Wheaton & Co. Ltd., Exeter

Foreword

The books in the Pergamon Thermodynamics and Fluid Mechanics Series were originally planned as a series for undergraduates, to cover those subjects taught in a three-year course for Mechanical Engineers. Subsequently, the aims of the series were broadened and several volumes were introduced which catered not only for undergraduates, but, also, for postgraduate students and engineers in practice. These included new editions of books published earlier in the series.

The present volume of worked problems by Mr Haywood is intended mainly for engineering students on undergraduate courses but it will also be useful for students on Masters degree courses in power plants. It aims to provide more self help for the reader.

With more importance being given to ENGINEERING APPLICATIONS in engineering education, as a result of the policies of the U.K. Engineering Council, the publication of this book is particularly appropriate and timely.

W. A. Woods
September 1986

Preface

During the nearly 20 years of continuing popularity of the author's Analysis of Engineering Cycles (Ref. 1), suggestions have been made from time to time that readers would appreciate having access to worked solutions to the Problems set at the end of each of the ten chapters, particularly since there are only a few worked examples in the body of the text. That latter fact was the result of deliberate policy on the part of the author, for two reasons. Firstly, it was considered that the insertion of too many worked examples would have hindered smooth development and presentation in the theoretical analysis of power, refrigerating and gas liquefaction plant. A second, but less important consideration, was the desirability of keeping the book to a reasonable size. Those considerations similarly influenced the author in his Equilibrium Thermodynamics for Engineers and Scientists (Ref. 3), to which reference is made herein.

It is appreciated that some take a different view in relation to worked examples in the text, particularly in the USA, and it is hoped that this companion volume to Ref. 1 will be particularly welcomed by these. They will have Prof. W. A. Woods to thank for persuading the author finally to succumb to requests from a number of sources for the publication of a volume of Worked Problems.

Full and detailed solutions are given for all the Problems set at the end of each chapter in the 3rd Edition of Ref 1. That edition contained all the problems that were set in the 2nd Edition, but with the addition of some ten new ones. Thus this volume will prove to be equally useful to those readers who presently only have access to the 2nd Edition, or to the Russian translation of that edition which is listed under Ref 1. The additional Problems in the 3rd Edition included coverage of types of plant which have assumed greater importance in recent years, namely Combined Heat and Power Plant (CHP) and Compressed Air Energy Storage Systems (CAES). They also allowed greater usage to be made of the important principles of thermodynamic availability and the concepts of lost work output (or greater work input) due to irreversibility. Readers will find adequate coverage of those principles and concepts in Appendix A of Ref 1., and a more exhaustive treatment in the author's other book (Ref 3).

In preparing this volume of Worked Problems, all the calculations have been completely reworked as a check on the accuracy of the quoted answers. The use of an electronic calculator inevitably led to very minor alterations to the last figure in some of the quoted answers. Only rarely was it found to be necessary to make a correction of any real significance. An error was discovered in the original framing of item (e) in Problem 9.4. That led to a consequential revision of items (e) and (f), and to the addition of a new item (g). In problem 10.9, items (a) and (b) have been interchanged, because it was found to be possible directly to calculate (b) first, without first calculating (a), as had been done originally.

The author's Thermodynamic Tables in SI (metric) Units (Ref 2) were used throughout the calculations.* In those few cases where additional data were required, the necessary information was given in the formal statement of the Problem in question.

I am particularly grateful to Mr J. E. Gilgunn-Jones, his staff at Pergamon Press and especially Mrs Sarah Purlan for coming to my rescue in producing camera-ready typescript from my hand-written manuscript, when my own arrangements for this fell through suddenly due to circumstances beyond my control.

R W HAYWOOD
September 1985

* Where occasional use was made of a more detailed set of Steam Tables (Ref.4), that fact has been noted in the text.

Contents

PART I

Simple Power and Refrigerating Plants

CHAPTER 1

Power Plant Performance Parameters

1.1 In a test of a cyclic steam power plant, the measured rate of steam supply was 7.1 kg/s when the net rate of work output was 5000 kW. The feed water was supplied to the boiler at a temperature of 38 °C, and the superheated steam leaving the boiler was at 1.4 MN/m² and 300 °C.

Calculate the thermal efficiency of the cycle and the corresponding heat rate. What would be the heat rate when expressed in Btu of heat input per kW h of work output?

Solution

$$\text{At } 1.4\ \text{MN/m}^2 \text{ and } 300\ °\text{C},\ h_s = 3041\ \text{kJ/kg}$$

$$\text{At } 38\ °\text{C},\ h_f = 159\ \text{kJ/kg}$$

$$\therefore Q_{in}/\text{kg} = 2882\ \text{kJ}$$

$$\therefore Q_{in} = 2882 \times 7.1 = 20460\ \text{kW}$$

$$\therefore \eta_{CY} \equiv \frac{W_{net}}{Q_{in}} = \frac{5000}{20460} \times 100 = 24.44,\ \text{say } \underline{24.4}\ \%$$

$$\text{Heat rate} = \frac{1}{0.2444} = \underline{4.09}$$

$$\text{But } 1\ \text{Btu} \approx 1.055\ \text{kJ} = 1.055\ \text{kW s}$$

$$\therefore 1\ \text{kW} = \frac{3600}{1.055} = 3412\ \text{Btu/h}$$

$$\therefore \text{Heat rate} = 4.09 \times 3412 = \underline{13960}\ \text{BTu/kW h}$$

1.2 In the test of the power plant of Problem 1.1, the coal supply rate was 3250 kg/h, and the calorific value of the coal was 26700 kJ/kg. Calculate the boiler efficiency, and the overall efficiency of the complete plant.

Solution

$$\text{Energy supplied in the coal} = \frac{3250 \times 26700}{3600} = 24100 \text{ kW}$$

$$\text{Boiler efficiency, } \eta_B \equiv \frac{Q_{in}}{CV} = \frac{20460}{24100} \times 100 = \underline{84.9} \ \%$$

$$\text{Overall efficiency, } \eta_o = \eta_{CY}\,\eta_B = 0.2444 \times 0.849 \times 100 = \underline{20.7} \ \%$$

1.3 A cyclic steam power plant is designed to supply steam from the boiler at 10 MN/m^2 and 550 °C when the boiler is supplied with feed water at a temperature of 209.8 °C. It is estimated that the thermal efficiency of the cycle will be 38.4 % when the net power output is 100 MW. Calculate the steam consumption rate. The enthalpy of the feed water may be taken as being equal to the enthalpy of saturated water at the same temperature.

The boiler has an estimated efficiency of 87 %, and the calorific value of the coal supplied is 25500 kJ/kg. Calculate the rate of coal consumption in (a) kg/s, (b) ton/min.

Solution

At 10 MN/m^2 and 550 °C, h_s = 3500 kJ/kg

At 209.8 °C, h_f = 897 kJ/kg

$\therefore Q_{in}/kg$ = 2603 kJ

Let the steam flow rate = $\dot{m}_s$ kg/s, for a power output of 100 MW

Then $\dot{m}_s Q_{in} \eta_{CY}$ = power output

$$\dot{m}_s \times 2603 \times 0.384 = 100 \times 10^3$$

$$\therefore \dot{m}_s = \underline{100} \text{ kg/s}$$

Let the coal supply rate = $\dot{m}_c$ kg/s, for a power output of 100 MW

Then $\dot{m}_c\, CV\, \eta_B\, \eta_{CY}$ = power output

$$\dot{m}_c \times 25500 \times 0.87 \times 0.384 = 100 \times 10^3$$

$$\therefore \dot{m}_c = \underline{11.74} \text{ kg/s}$$

$$= \frac{11.74 \times 60}{0.4536 \times 2240} = \underline{0.693} \text{ ton/min}$$

1.4 Air flows around a closed-circuit gas turbine plant, entering the compressor at 18 °C and leaving it at 190 °C. The air temperature is 730 °C at turbine inlet and 450 °C at turbine outlet. The values of the mean specific heat capacity of air over the temperature ranges occurring in the compressor, heater and turbine are respectively 1.01, 1.08 and 1.11 kJ/kg K. Calculate the cycle efficiency, neglecting mechanical losses.

Solution

$$W_T = 1.11 \times (730 - 450) = 310.8 \text{ kJ/kg}$$

$$W_C = 1.01 \times (190 - 18) = 173.7 \text{ kJ/kg}$$

$$\therefore W_{net} = (W_T - W_C) = 137.1 \text{ kJ/kg}$$

$$Q_{in} = 1.08 \times (730 - 190) = 583.2 \text{ kJ/kg}$$

$$\therefore \eta_{CY} \equiv \frac{W_{net}}{Q_{in}} = \frac{137.1}{583.2} \times 100 = \underline{23.5}\ \%$$

1.5 Calculate the required rate of air circulation per second round the circuit of Problem 1.4 if the net power output is to be 5000 kW. Calculate also the required rate of fuel supply when oil of calorific value 44500 kJ/kg is supplied to the combustion chamber, the efficiency of which as a heating device is 75 %.

Solution

Let the air flow rate = $\dot{m}_a$ kg/s, for a power output of 5000 kW

$$\text{Then } \dot{m}_a = \frac{5000}{137.1} = 36.47, \text{ say } \underline{36.5} \text{ kg/s}$$

Let the fuel supply rate = $\dot{m}_f$ kg/s

$$\text{Then } \dot{m}_f \, CV \, \eta_B = \dot{m}_a \, Q_{in}$$

$$\dot{m}_f \times 44500 \times 0.75 = 36.47 \times 583.2$$

$$\therefore \dot{m}_f = \underline{0.637} \text{ kg/s}$$

1.6 In the plant shown diagrammatically in Fig. 1.4, fuel and air at the pressure and temperature of the environment, namely 1 atm and 25 °C, are fed to an imperfectly lagged combustion chamber. The products of combustion then pass through a heat exchanger in which they transfer heat to the fluid of a cyclic heat power plant and finally exhaust to the atmosphere. Stray heat loss from the heat exchanger is negligible.

In a test on the plant, in which W_{net} was 11 MJ per kg of fuel burnt, the values of enthalpy and entropy at the points indicated were found to be as listed in the following table, all quantities being expressed per kg of fuel supplied:

State point	1	2	3	Products at 1 atm and 25 °C
Enthalpy, MJ	61.00	58.50	22.63	10.99
Entropy, MJ/K	0.2398	0.3070	0.2698	0.2429

Calculate the following quantities:

(a) The stray heat loss, Q_{12}, to the environment from the combustion chamber, expressed as a percentage of the calorific value of the fuel.

(b) The heating-device efficiency of the combined combustion chamber and heat exchanger.

(c) The thermal efficiency of the cyclic heat power plant.

(d) The thermal efficiency of an ideal, fully reversible, cyclic heat power plant which takes in heat by cooling the products of combustion between the same states 2 and 3 and in which heat rejection takes place at the temperature of the environment.

(e) The overall efficiency of the complete plant.

(f) The rational efficiency of the complete plant.

(g) The lost work due to irreversibility (1) in the combustion-chamber process and (2) in consequence of the discharge of the gases from the plant at a temperature in excess of the environment temperature.

Note: Before attempting (f) and (g), study Appendix A in Ref. 1.

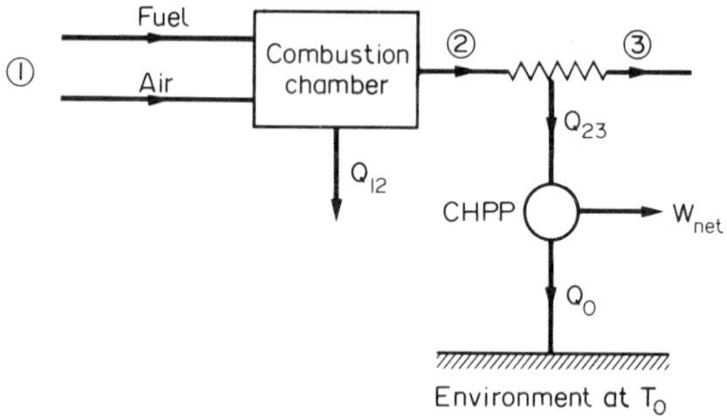

Fig. P.1.6 (Fig. 1.4)

Solution

(a) $$\frac{Q_{12}}{CV} = \frac{H_1 - H_2}{H_1 - H_0} = \frac{61.00 - 58.50}{61.00 - 10.99} = \frac{2.50}{50.01} \times 100 = \underline{5.0}\ \%$$

(where $H_0 \equiv$ enthalpy of products at 1 atm and 25 °C)

(b) $$\eta_B = \frac{H_1 - H_3}{CV} = \frac{61.00 - 22.63}{50.01} = \frac{38.37}{50.01} \times 100 = \underline{76.7}\ \%$$

(c) $$\eta_{CY} = \frac{W_{net}}{Q_{23}} = \frac{W_{net}}{H_2 - H_3} = \frac{11}{58.50 - 22.63} = \frac{11}{35.87} \times 100$$

$$= \underline{30.7}\ \%$$

(d) Ideal cycle $W_{net} \equiv [(W_X)_{REV}]_2^3 = B_2 - B_3$, where $B \equiv H - T_0S$

$= (H_2 - H_3) - T_0(S_2 - S_3)$

$= 35.87 - 298.15\ (0.3070 - 0.2698)$

$= 35.87 - 11.09 = 24.78$ MJ

$$\text{Ideal } \eta_{CY} \equiv \frac{\text{Ideal } W_{net}}{Q_{23}} = \frac{24.78}{35.87} \times 100 = \underline{69.1}\ \%$$

(e) $\eta_o \equiv \frac{W_{net}}{CV} = \frac{11}{50.01} \times 100 = \underline{22.0}\ \%$

(f) Ideal plant $W_{net} \equiv [(W_X)_{REV}]_1^0 = B_1 - B_0$

$= (H_1 - H_0) - T_0(S_1 - S_0)$

$= 50.01 - 298.15\ (0.2398 - 0.2429)$

$= 50.01 + 0.92 = 50.93$ MJ

$$\text{Rational efficiency, } \eta_R \equiv \frac{W_{net}}{\text{Ideal plant } W_{net}}$$

$$= \frac{11}{50.93} \times 100 = \underline{21.6}\ \%$$

(g) Lost work due to irreversibility $= T_0\ \Delta S_C$,

where $\Delta S_C \equiv$ entropy creation due to irreversibility

$= \Delta S - \frac{\Delta Q}{T}$, since here T is constant, being equal to T_0,

and $\Delta Q \equiv$ Heat <u>in</u> to control volume (i.e. the combustion chamber)

(1) <u>In the combustion chamber process</u>

$$\Delta S_C = (S_2 - S_1) - \frac{(-Q_{12})}{T_0}, \text{ where } Q_{12} = \text{stray heat } \underline{\text{loss}}.$$

$\therefore$ Lost work $= T_0(S_2 - S_1) + Q_{12}$

$= 298.15\ (0.3070 - 0.2398) + 2.50$

$= 20.03 + 2.50 = \underline{22.53}$ MJ

(2) <u>Due to the excess temperature of the exhaust gases</u>

Lost work $= (B_3 - B_0)$

$= (H_3 - H_0) - T_0(S_3 - S_0)$

$= (22.63 - 10.99) - 298.15\ (0.2698 - 0.2429)$

$= 11.64 - 8.02 = \underline{3.62}$ MJ

CHAPTER 2

Simple Steam Plant

2.1 For the conditions given in Example 2.1, express the Rankine cycle efficiency as a percentage of the Carnot cycle efficiency for the same extreme limits of temperature.

Solution

From Example 2.1, $\eta_{RANK} = 32.80$ %

With reference to Fig. 2.1:

$$\eta_{CARNOT} = \frac{T_3 - T_1}{T_1}$$

$$\text{where } T_3 = 350 + 273.15 = 623.15 \text{ K}$$

$$\text{and } T_1 = 39.0 + 273.15 = 312.15 \text{ K}$$

$$\therefore \eta_{CARNOT} = \frac{311}{623.15} \times 100 = 49.91 \text{ \%}$$

$$\therefore \frac{\eta_{RANK}}{\eta_{CARNOT}} = \frac{32.80}{49.91} \times 100 = \underline{65.7} \text{ \%}$$

2.2 Calculate the thermal efficiency of a Rankine cycle in which the steam is initially dry saturated at 2 MN/m^2 and the condenser pressure is 7 kN/m^2. Express this efficiency as a percentage of the Carnot cycle efficiency for the same limits of temperature. Explain why the Rankine cycle efficiency is less than in Example 2.1, and why the ratio of the efficiencies of the respective Rankine and Carnot cycles is greater.

Solution

Neglecting the feed pump work input, and using the notation of Fig. 2.1:

$$h_3 = 2797.2 \text{ kJ/kg}, \qquad h_7 = 2572.6 \text{ kJ/kg}$$

$$s_4 = s_3 = 6.337 \text{ kJ/kg K}, \qquad s_7 = 8.277 \text{ kJ/kg K}$$

$$\therefore \quad h_4 = h_7 - T_1(s_7 - s_4)$$

$$= 2572.6 - 312.15\ (8.277 - 6.337)$$

$$= 1967.0 \text{ kJ/kg}$$

$$W_T = h_3 - h_4 = 2797.2 - 1967.0 = 830.2 \text{ kJ/kg}$$

$$Q_{in} \approx h_3 - h_1 = 2797.2 - 163.4 = 2633.8 \text{ kJ/kg}$$

Neglecting the effects of the feed pump,

$$\eta_{RANK} = \frac{W_T}{Q_{in}} = \frac{830.2}{2633.8} \times 100 = 31.52, \text{ say } \underline{\mathit{31.5}}\ \%$$

$$\eta_{CARNOT} = \frac{T_3 - T_1}{T_1},$$

$$\text{where } T_3 = 212.4 + 273.15 = 485.55 \text{ K}$$

$$T_1 = 39.0 + 273.15 = 312.15 \text{ K}$$

$$\therefore \quad \eta_{CARNOT} = \frac{173.4}{485.55} \times 100 = 35.71\ \%$$

$$\therefore \quad \frac{\eta_{RANK}}{\eta_{CARNOT}} = \frac{31.52}{35.71} \times 100 = \underline{\mathit{88.3}}\ \%$$

η_{RANK} in this Problem (with dry saturated steam) is less than in Example 2.1 (with superheated steam) because the mean temperature of heat reception here is less than that in Example 2.1.

$\eta_{RANK}/\eta_{CARNOT}$ is greater with dry saturated steam than with superheated steam because, with the former, the mean temperature of heat reception (see Section 7.1 of Chapter 7) is much closer to the top temperature in the cycle than it is with superheated steam. Thus, with dry saturated steam, η_{RANK} more closely approaches the thermal efficiency of a Carnot cycle which has the same top temperature as the respective Rankine cycle. (The reader should draw the respective temperature-entropy diagrams and indicate on each the mean temperature of heat reception, as in Fig. 7.1 of Chapter 7.)

2.3 In a steam power plant the turbine has an isentropic efficiency of 80 % when the conditions are otherwise as in Problem 2.2. Neglecting the work input to the feed pump, determine the thermal efficiency of the plant if the condensate is returned to the boiler at the saturation temperature corresponding to the condenser pressure. Calculate the steam consumption rate, in kg per MJ of turbine work output.

If the steam is supplied from a boiler of 84 % efficiency, and the calorific value of the coal is 28000 kJ/kg, calculate the overall efficiency of the plant and the coal consumption rate in kg/MJ of turbine work output.

Solution

From Problem 2.2, isentropic $W_T = 830.2$ kJ/kg

$\therefore$ Actual $W_T = 0.8 \times 830.2 = 664.2$ kJ/kg

From Problem 2.2, $Q_{in} = 2633.8$ kJ/kg

$$\therefore \eta_{CY} = \frac{664.2}{2633.8} \times 100 = 25.22,$$

say *25.2* %

Let the steam consumption rate $= m_s$ kg/MJ

Then, for a turbine work output of 1 MJ (= 1000 kJ),

$$m_s \times 664.2 = 1000$$

$$\therefore m_s = \underline{1.506} \text{ kg/MJ}$$

Overall efficiency, $\eta_o = \eta_B \, \eta_{CY} = 0.84 \times 25.22 = 21.18$, say *21.2* %

Let the coal consumption rate $= m_c$ kg/MJ

Then $m_c \, CV \, \eta_o$ = work output

$$m_c \times 28000 \times 0.2118 = 1000$$

$$\therefore m_c = \underline{0.169} \text{ kg/MJ}$$

2.4 For the plant of Problem 1.1, determine the efficiency ratio, as defined in Section 2.5, and the rational efficiency of the work-producing steam circuit, as calculated from eqn. (2.9) of Section 2.8. Take the environment temperature as being equal to the saturation temperature of the steam in the condenser, namely 38 °C.

Solution

$$\text{By definition, efficiency ratio} \equiv \frac{\text{Actual } \eta_{CY}}{\eta_{RANK}}$$

We shall here calculate the exact Rankine cycle efficiency taking into account the effects of the feed pump, and using the notation of Fig. 2.1.

At 1.4 MN/m² and 300 °C, $h_3 = 3041$ kJ/kg, $s_3 = 6.983$ kJ/kg K

To calculate the dryness fraction, x_4, at state point 4 (at 38 °C):

$$(1 - x_4) = \frac{s_7 - s_4}{s_7 - s_1}, \text{ where } s_7 = 8.296 \text{ kJ/kg K}$$

$$\text{and } s_1 = 0.545 \text{ kJ/kg K}$$

$$\text{But } s_4 = s_3 = 6.983 \text{ kJ/kg K}$$

$$\therefore (1 - x_4) = \frac{8.296 - 6.983}{8.296 - 0.545} = 0.1694$$

$$h_4 = h_7 - (1 - x_4)(h_7 - h_1)$$

$$\text{where } h_7 = 2570.8 \text{ kJ/kg,}$$

$$(h_7 - h_1) = h_{fg} = 2411.7 \text{ kJ/kg}$$

$$\therefore h_4 = 2570.8 - 0.1694 \times 2411.7 = 2162.3 \text{ kJ/kg}$$

$$\therefore \text{Turbine work output, } W_T = (h_3 - h_4) = 3041 - 2162.3$$

$$= 878.7 \text{ kJ/kg}$$

$$\text{Feed pump work input, } W_P = \int_1^2 v\,dp = v_1 (p_B - p_A)$$

$$\text{where } v_1 = 0.001007 \text{ m}^3\text{/kg, } p_B = 1400 \text{ kN/m}^2,$$

$$p_A = 6.62 \text{ kN/m}^2$$

$$\therefore W_P = 0.001007 \times 1393.38 = 1.4 \text{ kJ/kg}$$

$$W_{net} = W_T - W_P = 878.7 - 1.4 = 877.3 \text{ kJ/kg}$$

$$Q_{in} = (h_3 - h_2),$$

$$\text{where } h_2 = h_1 + W_P = 159.1 + 1.4 = 160.5 \text{ kJ/kg}$$

$$\therefore Q_{in} = 3041 - 160.5 = 2880.5 \text{ kJ/kg}$$

$$\text{Exact } \eta_{RANK} \equiv \frac{W_{net}}{Q_{in}} = \frac{877.3}{2880.5} \times 100 = 30.46\ \%$$

$$\text{Actual } \eta_{CY} \equiv \frac{\text{Actual } W_{net}}{\text{Actual } Q_{in}},$$

$$\text{where Actual } W_{net} = 5000 \text{ kW}$$

$$\text{and Actual } Q_{in} = \dot{m}_s (h_3 - h_2) = 7.1 \times 2880.5 = 20452 \text{ kW}$$

$$\therefore \text{Actual } \eta_{CY} = \frac{5000}{20452} \times 100 = 24.45\ \%$$

$$\text{Efficiency ratio} \equiv \frac{\text{Actual } \eta_{CY}}{\eta_{RANK}} = \frac{24.45}{30.46} = \underline{0.803}$$

$$\text{By definition, } (\eta_R) \text{ steam circuit} \equiv \frac{\text{Actual } W_{net}}{\text{Ideal } W_{net}}$$

$$\text{where Ideal } W_{net} = (b_3 - b_2)$$

$$= (h_3 - h_2) - T_0(s_3 - s_2)$$

$$s_3 = 6.983 \text{ kJ/kg K and } T_0 = 38 + 273.15 = 311.15 \text{ K}$$

$$s_2 = s_1 = s_f = 0.545 \text{ kJ/kg K}$$

$$\therefore \text{ Ideal } W_{net} = 2880.5 - 311.15 \times 6.438$$

$$= 877.3 \text{ kJ/kg}$$

$$\text{Actual } W_{net} = \frac{5000}{7.1} = 704.2 \text{ kJ/kg}$$

$$\therefore (\eta_R)_{\text{steam circuit}} = \frac{704.2}{877.3} \times 100 = \underline{80.3} \text{ \%}$$

Note: Comparing this with the calculated value of the efficiency ratio, it is seen that this ratio is numerically equal to the rational efficiency of the steam circuit, namely of the open-circuit work-producing plant within control surface S of Fig. 2.3. Comparing the lengths of the two calculations, it is also seen that, as pointed out in Section 2.8, it is simpler to obtain the value of the efficiency ratio by calculating directly the value of the rational efficiency of the steam circuit.

2.5 For the plant of Problem 2.3, write down the values of the efficiency ratio and the rational efficiency of the work-producing steam circuit.

Solution

$$\text{Efficiency ratio} \equiv \frac{\text{Actual } \eta_{CY}}{\text{Rankine } \eta_{CY}} = \frac{\text{Actual } W_{net}}{\text{Rankine } W_{net}}$$

$$\approx \frac{\text{Actual } W_T}{\text{Rankine } W_T} \text{ (neglecting feed pump work input)}$$

$$= \text{Turbine isentropic efficiency (80 \%)}$$

$$\therefore \text{ Efficiency ratio} = \underline{0.8}$$

$$\text{and } (\eta_R)_{\text{steam circuit}} = \underline{80} \text{ \%}$$

2.6 Steam is expanded isentropically in the high-pressure cylinder of a steam turbine from a pressure of 2 MN/m² at 350 °C to 0.1 MN/m². Isentropic expansion is then continued down to 7 kN/m² in the low-pressure cylinder. Calculate the percentage of the total work output that is performed by the HP cylinder. (See Example 2.1 for the total work output of both cylinders.)

Solution

Using the notation of Fig. 2.1, and additionally describing the condition of the steam at exhaust from the HP cylinder as state point 3' (after isentropic expansion to 0.1 MN/m² from state point 3):

At 2 MN/m² and 350 °C, h_3 = 3139 kJ/kg, s_3 = 6.960 kJ/kg K

At HP cylinder exhaust, pressure = 0.1 MN/m² (at which s_g = 7.360 kJ/kg K)

But $s_{3'} = s_3 = 6.960$ kJ/kg K, so $s_{3'} < s_g$ and the steam is wet at 3'

$T_{3'} = 99.6 + 273.15 = 372.75$ K

$h_{3'} = h_g - T_{3'}(s_g - s_{3'})$, where g relates to dry saturated steam at 0.1 MN/m²

$= 2675.4 - 372.75\ (7.360 - 6.960) = 2526.3$ kJ/kg

$\therefore$ Work output from HP cylinder $= h_3 - h_{3'}$

$= 3139 - 2526.3$

$= 612.7$ kJ/kg

From Example 2.1:

Total work output $= W_T = 977.2$ kJ/kg

$\therefore$ Percentage in HP cylinder $= \dfrac{612.7}{977.2} \times 100 = \underline{62.7\ \%}$

2.7 The expansion in a turbine is adiabatic and irreversible. The entropy of the steam at inlet is 6.939 kJ/kg K, and the turbine exhausts at a pressure of 7 kN/m². If the dryness fraction of the steam at exhaust is 0.91, calculate the lost work due to irreversibility per kilogram of steam flowing through the turbine. If the inlet pressure is 4 MN/m², what is the isentropic efficiency of the turbine?

Solution

Let state 1 be the steam condition at turbine inlet and state 2 that at turbine exhaust.

Let the condition of the steam after isentropic expansion from state 1 to the pressure at exhaust be described as state 2_s.

At state 2 (turbine exhaust), dryness fraction = 0.91

$s_2 = s_g - (1 - x_2)(s_g - s_f)$ at 7 kN/m²

$= 8.277 - 0.09\ (8.277 - 0.559)$

$= 7.582$ kJ/kg K

At state 1 (turbine inlet), $s_1 = 6.939$ kJ/kg K (given)

Entropy creation in the turbine, due to irreversibility,

$\Delta S_c = (s_2 - s_1)$

$\therefore \Delta S_c = 7.582 - 6.939 = 0.643$ kJ/kg K

Lost work due to irreversibility $= T_0 \Delta S_c$,

where T_0 = environment temperature

Taking the environment temperature to be the same as the temperature of the steam at turbine exhaust, namely 39.0 °C (at 7 kN/m²)

$$T_0 = 39.0 + 273.15 = 312.15 \text{ K}$$

$$\therefore \text{ Lost work due to irreversibility} = 312.15 \times 0.643$$

$$= \underline{200.7} \text{ kJ/kg}$$

At turbine inlet: $p_1 = 4$ MN/m², $s_1 = 6.939$ kJ/kg K (given)

$$\therefore t_1 = 450 \text{ °C}, \ h_1 = 3331 \text{ kJ/kg}$$

At turbine exhaust: $p_2 = 7$ kN/m², $h_g = 2572.6$ kJ/kg,

$$h_{fg} = 2409.2 \text{ kJ/kg}$$

$$h_2 = h_g - (1 - x_2)h_{fg}$$

$$= 2572.6 - 0.09 \times 2409.2$$

$$= 2355.8 \text{ kJ/kg}$$

In the turbine:

Actual enthalpy drop, $\Delta h = (h_1 - h_2) = 3331 - 2355.8$

$$= 975.2 \text{ kJ/kg}$$

Isentropic enthalpy drop, $\Delta h_s = (h_1 - h_{2_s})$

$$= (h_1 - h_2) + (h_2 - h_{2_s})$$

$$= \Delta h + T_0(s_2 - s_{2_s})$$

But $s_{2_s} = s_1$

$$\therefore \Delta h_s = \Delta h + T_0 \Delta S_c$$

$$= 975.2 + 200.7 = 1175.9 \text{ kJ/kg}$$

Turbine isentropic efficiency, $\eta_T \equiv \dfrac{\Delta h}{\Delta h_s} = \dfrac{975.2}{1175.9} \times 100$

$$= \underline{82.9} \ \%$$

CHAPTER 3

Simple Closed-Circuit Gas-Turbine Plant

In these Problems, air is to be treated as a perfect gas with $\gamma = 1.4$ and $c_p = 1.01$ kJ/kg K.

CLOSED-CIRCUIT PLANT

3.1 In an air-standard Joule cycle the temperatures at compressor inlet and outlet are respectively 60 °C and 170 °C, and the temperature at turbine inlet is 600 °C.

Calculate (a) the temperature at turbine exhaust; (b) the turbine work and compressor work per kg of air; (c) the thermal efficiency of the cycle; (d) the pressure ratio.

Solution

Using the notation of Fig. 3.1, $T_2/T_1 = T_3/T_4$.

(a) $T_4 = T_3 \times (T_1/T_2) = 873.15 \times (333.15/443.15) = 656.41$ K

$\therefore\ t_4 = \underline{383.3}$ °C

(b) $W_T = c_p(T_3 - T_4) = 1.01\ (600 - 383.3) = \underline{218.9}$ kJ/kg

$W_C = c_p(T_2 - T_1) = 1.01\ (170 - 60) = \underline{111.1}$ kJ/kg

(c) $Q_{in} = c_p(T_3 - T_2) = 1.01\ (600 - 170) = 434.3$ kJ/kg

$$\eta_{CY} \equiv \frac{W_{net}}{Q_{in}} = \frac{(218.9 - 111.1)}{434.3} \times 100 = \underline{24.8}\ \%$$

(d) Isentropic temperature ratio, $\rho_p = r_p^{(\gamma-1)/\gamma}$, where r_p = pressure ratio

$$\therefore\ r_p^{0.4/1.4} = \frac{443.15}{333.15} = 1.330$$

$$\therefore\ r_p = (1.330)^{3.5} = \underline{2.71}$$

3.2 In a closed-circuit gas-turbine plant using a perfect gas as the working fluid, the thermodynamic temperatures at compressor and turbine inlets are respectively T_a and T_b. The plant is operating with an isentropic temperature ratio of compression of ρ_p and the isentropic efficiencies of the compressor and turbine are respectively η_C and η_T. Show that the ratio of the compressor work input to the turbine work output is given by

$$\frac{W_C}{W_T} = \frac{\rho_p}{\alpha}, \text{ where } \alpha \equiv \eta_C\ \eta_T\ \theta \text{ and } \theta \equiv T_b/T_a.$$

Evaluate this ratio when t_a = 20 °C, t_b = 700 °C, $\eta_C = \eta_T$ = 85 %, the pressure ratio = 4.13 and the working fluid is air.

Solution

Using the notation of Fig. 3.4, and the equations for W_C and W_T given in Section 3.7:

$$W_C = c_p\ T_a\ (\rho_p - 1)/\eta_C \tag{3.8}$$

$$W_T = c_p\ \eta_T T_b (1 - 1/\rho_p) \tag{3.9}$$

$$\therefore\ \frac{W_C}{W_T} = \frac{\rho_p\ T_a}{\eta_C\ \eta_T\ T_b} = \underline{\underline{\frac{\rho_p}{\alpha}}}, \text{ where } \alpha \equiv \eta_C\ \eta_T\ \theta \text{ and } \theta \equiv T_b/T_a$$

$$T_b = 973.15\text{ K},\ T_a = 293.15\text{ K},\ \rho_p = r_p^{(\gamma-1)/\gamma} = 4.13^{1/3.5}$$
$$= 1.500$$

$$\alpha \equiv \eta_C\ \eta_T\ \frac{T_b}{T_a} = 0.85 \times 0.85 \times \frac{973.15}{293.15} = 2.398$$

$$\therefore\ \frac{W_C}{W_T} = \frac{\rho_p}{\alpha} = \frac{1.500}{2.398} = \underline{0.625}$$

3.3 For the plant described in Problem 3.2, show that the thermal efficiency of the cycle is given by

$$\eta_{CY} = \frac{(1 - 1/\rho_p)(\alpha - \rho_p)}{(\beta - \rho_p)},$$

where $\alpha \equiv \eta_C\ \eta_T\ \theta$,

$\beta \equiv [1 + \eta_C(\theta - 1)]$,

$\theta \equiv T_b/T_a$.

For the values given in Problem 3.2, calculate the thermal efficiency (a) using this expression, (b) by first calculating the temperatures at outlet from the compressor and turbine.

Solution

From equations (3.8) and (3.9)

$$W_{net} = W_T - W_C = [c_p \eta_T T_b (1 - 1/\rho_p)] - [c_p T_a (\rho_p - 1)/\eta_C]$$

$$\therefore W_{net} = c_p T_a (1 - 1/\rho_p)(\eta_T \theta - \rho_p/\eta_C),$$

where $\theta \equiv T_b/T_a$

$$= \frac{c_p T_a}{\eta_C}(1 - 1/\rho_p)(\eta_C \eta_T \theta - \rho_p)$$

$$\therefore W_{net} = \frac{c_p T_a}{\eta_C}(1 - 1/\rho_p)(\alpha - \rho_p), \text{ where } \alpha \equiv \eta_C \eta_T \theta$$

$$\text{Now } (T_{2'} - T_a) = (T_2 - T_a)/\eta_C = T_a (\rho_p - 1)/\eta_C$$

$$\therefore T_{2'} = T_a [1 + (\rho_p - 1)/\eta_C],$$

$$\text{and } Q_{in} = c_p (T_b - T_{2'}) = \frac{c_p T_a}{\eta_C}\left[\eta_C \frac{T_b}{T_a} - (\eta_C + \rho_p - 1)\right]$$

$$= \frac{c_p T_a}{\eta_C}(\eta_C \theta - \eta_C - \rho_p + 1)$$

$$= \frac{c_p T_a}{\eta_C}[1 + \eta_C(\theta - 1) - \rho_p]$$

$$\therefore Q_{in} = \frac{c_p T_a}{\eta_C}(\beta - \rho_p),$$

where $\beta \equiv [1 + \eta_C(\theta - 1)]$

$$\text{whence } \eta_{CY} \equiv W_{net}/Q_{in} = \frac{(1 - 1/\rho_p)(\alpha - \rho_p)}{(\beta - \rho_p)}$$

(a) From Problem 3.2, $\rho_p = 1.5 \quad \therefore (1 - 1/\rho_p) = \frac{1}{3}$

$$\theta \equiv T_b/T_a = 973.15/293.15 = 3.3196$$

$$\alpha \equiv \eta_C \eta_T \theta = 0.85 \times 0.85 \times 3.3196 = 2.398$$

$$\beta \equiv [1 + \eta_C(\theta - 1)] = [1 + 0.85 \times 2.3196]$$

$$= 2.972$$

$$(\alpha - \rho_p) = 0.898$$

$$(\beta - \rho_p) = 1.472$$

$$\therefore \eta_{CY} = \frac{1}{3} \times \frac{0.898}{1.472} \times 100 = \underline{20.3}\ \%$$

(b) Using the notation of Fig. 3.4:

$$\frac{T_2}{T_a} = \rho_p = 1.500 \qquad \therefore (T_2 - T_a) = 0.5\ T_a$$

$$(T_{2'} - T_a) = \frac{T_2 - T_a}{\eta_C} = \frac{0.5\ T_a}{\eta_C} = \frac{0.5 \times 293.15}{0.85} = 172.44$$

$\therefore$ At compressor outlet, $T_{2'} = 172.44 + 293.15 = 465.59$ K

$$\frac{T_b}{T_4} = \rho_p = 1.500 \quad \therefore (T_b - T_4) = T_b\ (1 - \frac{1}{\rho_p}) = \frac{T_b}{3} = \frac{973.15}{3} = 324.38$$

$$(T_b - T_{4'}) = \eta_T\ (T_b - T_4) = 0.85 \times 324.38 = 275.72$$

$\therefore$ At turbine outlet, $T_{4'} = 973.15 - 275.72 = 697.43$ K

$$W_T = c_p\ (T_b - T_{4'}) = 275.72\ c_p$$

$$W_C = c_p\ (T_{2'} - T_a) = 172.44\ c_p$$

$$\therefore W_{net} = W_T - W_C = 103.28\ c_p$$

$$Q_{in} = c_p\ (T_b - T_{2'}) = c_p\ (973.15 - 465.59) = 507.56\ c_p$$

$$\therefore \eta_{CY} \equiv \frac{W_{net}}{Q_{in}} = \frac{103.28\ c_p}{507.56\ c_p} \times 100 = \underline{20.3}\ \%$$

3.4 In a design study for the plant described in Problem 3.2, T_a, T_b, η_C and η_T are kept constant while the pressure ratio of compression for which the plant is to be designed is varied.

Show that, when the design pressure ratio is changed by a small amount, the changes in heat rejected and heat supplied are related by the expression

$$\frac{\delta Q_{out}}{\delta Q_{in}} = \frac{\alpha}{\rho_p^2}.$$

Also show that W_{net} has its maximum value when $\delta Q_{out} = \delta Q_{in}$, while η_{CY} has its maximum value η_m when

$$\frac{\delta Q_{out}}{Q_{out}} = \frac{\delta Q_{in}}{Q_{in}}.$$

Hence show that, if ρ_w and $\rho_{opt.}$ are the values of ρ_p for maximum W_{net} and maximum η_{CY} respectively, then

$$\frac{\rho_w}{\rho_{opt.}} = \sqrt{(1 - \eta_m)}.$$

Solution

Using the notation of Fig. 3.4:

$$Q_{in}/c_p = (T_b - T_{2'}) = T_b - T_a\,[1 + (\rho_p - 1)/\eta_C]$$

$$\therefore \frac{1}{c_p}\frac{\delta Q_{in}}{\delta \rho_p} = -\frac{T_a}{\eta_C} \qquad (1)$$

$$Q_{out}/c_p = (T_{4'} - T_a) = [T_b - \eta_T(T_b - T_4)] - T_a$$

$$= T_b[1 - \eta_T(1 - 1/\rho_p)] - T_a$$

$$\therefore \frac{1}{c_p}\frac{\delta Q_{out}}{\delta \rho_p} = -\frac{\eta_T\,T_b}{\rho_p^2} \qquad (2)$$

From (1) and (2)

$$\frac{\delta Q_{out}}{\delta Q_{in}} = \frac{\eta_C\,\eta_T\,T_b/T_a}{\rho_p^2} = \frac{\eta_C\,\eta_T\,\theta}{\rho_p^2} = \underline{\underline{\frac{\alpha}{\rho_p^2}}} \qquad (3)$$

From Problem 3.3:

$$W_{net} = \frac{c_p\,T_a}{\eta_C}(1 - 1/\rho_p)(\alpha - \rho_p)$$

$$\therefore \frac{\delta W_{net}}{\delta \rho_p} = \frac{c_p\,T_a}{\eta_C}\left[\frac{\alpha - \rho_p}{\rho_p^2} - (1 - \frac{1}{\rho_p})\right]$$

$$= \frac{c_p\,T_a}{\eta_C}\left(\frac{\alpha}{\rho_p^2} - 1\right) = 0 \text{ for maximum } W_{net}$$

$\therefore$ *W_{net} is a maximum* when $\alpha/\rho_p^2 = 1$ (4)

i.e. *when* $\underline{\underline{\delta Q_{out} = \delta Q_{in}}}$ [from (3)]

$$\eta_{CY} = 1 - \frac{Q_{out}}{Q_{in}} \equiv 1 - y, \text{ where } y \equiv \frac{Q_{out}}{Q_{in}}$$

η_{CY} is a maximum when y is a minimum

$$\log y = \log Q_{out} - \log Q_{in}$$

$$\therefore \frac{\delta y}{y} = \frac{\delta Q_{out}}{Q_{out}} - \frac{\delta Q_{in}}{Q_{in}} = 0 \text{ for maximum } \eta_{CY}$$

Hence, *for maximum* η_{CY}, $\dfrac{\delta Q_{out}}{Q_{out}} = \dfrac{\delta Q_{in}}{Q_{in}}$

Whence, from (3), $\dfrac{Q_{out}}{Q_{in}} = \dfrac{\delta Q_{out}}{\delta Q_{in}} = \dfrac{\alpha}{\rho^2_{opt.}}$

where $\rho_{opt.}$ is the value of ρ_p at maximum η_{CY}

$$\therefore \text{Maximum cycle efficiency, } \eta_m = (1 - \frac{Q_{out}}{Q_{in}}) = 1 - \frac{\alpha}{\rho^2_{opt.}} \quad (5)$$

At maximum W_{net}, $\rho_p \equiv \rho_w$ and $\alpha/\rho_p^2 = 1$ [from (4)]

$$\therefore \alpha = \rho_w^2 \quad (6)$$

Hence, from (5) and (6), $\eta_m = 1 - \dfrac{\rho_w^2}{\rho^2_{opt.}}$

$$\therefore \frac{\rho_w}{\rho_{opt.}} = \sqrt{1 - \eta_m}$$

OPEN-CIRCUIT PLANT

3.5 In a method of providing air for the cabin of a piston-engined aircraft flying at high altitude, the air supply is taken from the engine supercharger at a pressure of 1.17 bar and a temperature of 55 °C, and is then further compressed in a centrifugal compressor. The compressed air is then passed through an air cooler, in which it is cooled by atmospheric air before entering a turbine, which drives the centrifugal compressor. The air from the turbine exhaust is supplied to the cabin at a pressure of 0.86 bar. The isentropic efficiencies of the turbine and compressor are each 75 %. Mechanical losses, and pressure drops in the ducting and air cooler, may be neglected.

Sketch a temperature-entropy diagram for the air in its passage through the plant. If the air is to be supplied to the cabin at a temperature of 15 °C, show that the compressor must be designed for a pressure ratio of about 1.35. What is then the required temperature at turbine inlet, and how much heat must be transferred in the air cooler per kg of air flowing through it?

Solution

To show that, when the compressor pressure ratio = 1.35,

(a) Temperature at turbine exhaust = 15 °C when temperature from supercharger = 55 °C.

(b) $W_T = W_C$, as required.

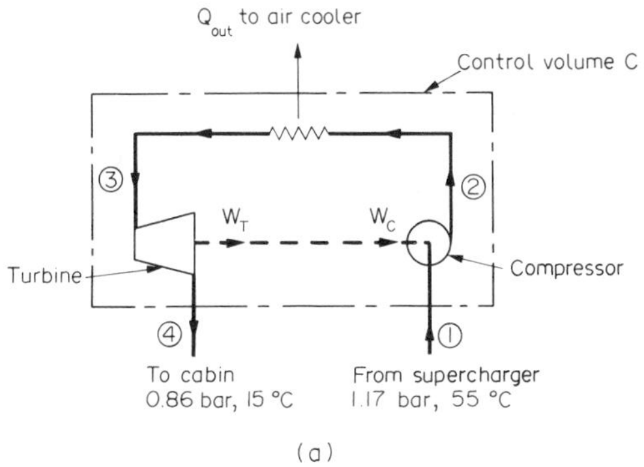

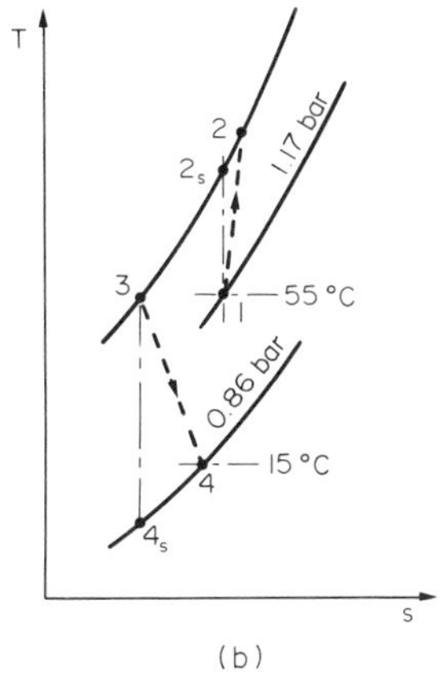

Fig. P.3.5 (a)   Fig. P.3.5 (b)

(a) Isentropic temperature ratio $\rho = r_p^{(\gamma-1)/\gamma}$, where r_p = pressure ratio

$$\gamma = 1.4 \quad \therefore \frac{\gamma - 1}{\gamma} = \frac{0.4}{1.4} = \frac{1}{3.5}$$

<u>In the compressor</u>, $\rho_c = (\frac{p_2}{p_1})^{\frac{\gamma-1}{\gamma}} = (1.35)^{\frac{1}{3.5}} = 1.0895$

<u>In the turbine</u>, $\rho_T = (\frac{p_3}{p_4})^{\frac{\gamma-1}{\gamma}} = (\frac{p_2}{p_1})^{\frac{\gamma-1}{\gamma}} (\frac{p_1}{p_4})^{\frac{\gamma-1}{\gamma}}$, since $p_3 = p_2$

$$= 1.0895 \times (\frac{1.17}{0.86})^{\frac{1}{3.5}}$$

$$= 1.0895 \times 1.0919 = 1.1896$$

<u>Compressor</u> $T_{2_s}/T_1 = \rho_c = 1.0895$, $T_1 = 328.15$ K

$$(T_{2_s} - T_1) = 0.0895 \times 328.15 = 29.37 \text{ K}$$

$$(T_2 - T_1) = \frac{T_{2_s} - T_1}{\eta_C} = \frac{29.37}{0.75} = 39.16 \text{ K}$$

$$W_C = c_p(T_2 - T_1) = 39.16\, c_p$$

and $T_2 = 328.15 + 39.16 = 367.31$ K

Air Cooler $$Q_{out} = c_p(T_2 - T_3)$$

But for the complete plant (control volume C),

$$Q_{out} = c_p(T_1 - T_4)$$

$$\text{Hence } T_2 - T_3 = T_1 - T_4 = 40$$

$$\therefore T_3 = 367.31 - 40 = 327.31 \text{ K}$$

$$\therefore t_3 = 327.31 - 273.15 = 54.16 \simeq \underline{54} \text{ °C}$$

Turbine

$$T_3/T_{4_s} = \rho_T = 1.1896$$

$$\therefore (T_3 - T_{4_s}) = T_3(1 - \frac{1}{\rho_T}) = 327.31\ (1 - \frac{1}{1.1896}) = 52.17$$

$$(T_3 - T_4) = \eta_T(T_3 - T_{4_s}) = 0.75 \times 52.17 = 39.13$$

$$\therefore t_4 = 54.16 - 39.13 = \underline{15.03} \text{ °C}$$

$$\text{But } t_4 = 15 \text{ °C (given)}$$

∴ *This condition is satisfied when* $r_p = 1.35$

(b) $$W_T = c_p(T_3 - T_4) = \underline{39.13\ c_p}$$

$$\text{But } W_C = \underline{39.16\ c_p}$$

∴ *This condition is also satisfied when* $r_p = 1.35$

Heat transferred in the air cooler

For the plant as a whole, within control volume C of Fig. P.3.5 (a):

$$Q_{out} \text{ to air cooler} = (h_1 - h_4) = c_p(T_1 - T_4)$$

$$= 1.01 \times 40 = \underline{40.4} \text{ kJ/kg}$$

3.6 A petrol engine is fitted with a turbo-supercharger, which comprises a centrifugal compressor driven by an exhaust gas-turbine. The gravimetric ratio of air to fuel supplied to the engine is 12, the fuel being mixed with the air between the compressor and the engine. The air is drawn into the compressor at a pressure of 1 bar and a temperature of 15 °C, and is delivered to the engine at a pressure of 1.5 bar. The exhaust gases from the engine enter the turbine at a pressure of 1.3 bar and a temperature of 510 °C and leave the turbine at a pressure of 1 bar. The isentropic efficiencies of the turbine and compressor are each 80 %.

Assuming that the thermal properties of the exhaust gases are the same as for air, calculate:

(a) the temperature of the air leaving the compressor;
(b) the temperature of the gases leaving the turbine;
(c) the power loss in the turbo-supercharger, due to external friction, expressed as a percentage of the power generated in the turbine.

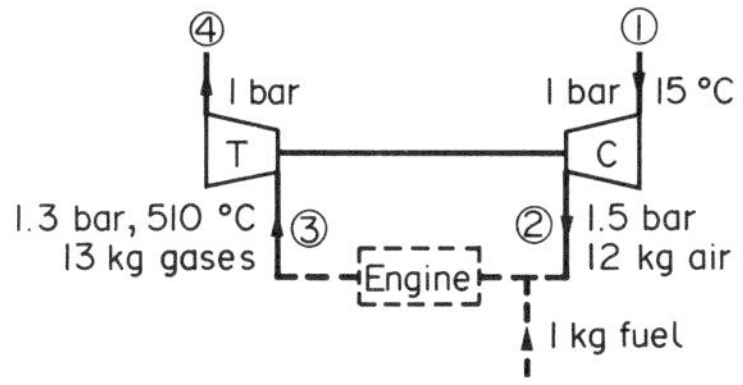

Fig. P.3.6.

Solution

(a) Compressor

$$\frac{T_{2_s}}{T_1} = \left(\frac{p_2}{p_1}\right)^{\frac{\gamma-1}{\gamma}} = (1.5)^{\frac{1}{3.5}} = 1.1228$$

$$(T_2 - T_1) = \frac{T_{2_s} - T_1}{\eta_C}$$

$$= \frac{0.1228 \times 288.15}{0.8}$$

$$= 44.2 \text{ K}$$

$$\therefore t_2 = 15 + 44.2 = \underline{59.2} \text{ °C}$$

(b) Turbine

$$\frac{T_3}{T_{4_s}} = \left(\frac{p_3}{p_4}\right)^{\frac{\gamma-1}{\gamma}} = (1.3)^{\frac{1}{3.5}} = 1.0778, \quad T_3 = 783.15 \text{ K}$$

$$(T_3 - T_4) = \eta_T (T_3 - T_{4_s})$$

$$= 0.8 \times 783.15 \left(1 - \frac{1}{1.0778}\right) = 45.2$$

$$\therefore t_4 = 510 - 45.2 = \underline{464.8} \text{ °C}$$

(c) Per kg of fuel burned:

$$W_T = 13 \times c_p \times 45.2 = 587.6\ c_p$$

$$W_C = 12 \times c_p \times 44.2 = \underline{530.4\ c_p}$$

$$\therefore \text{ External mechanical losses} = 57.2\ c_p$$

$$\text{Percentage loss} = \frac{57.2}{587.6} \times 100 = \underline{9.7}\ \%$$

CHAPTER 4

Internal-Combustion Power Plant

4.1 Calculate the temperature rise during reversible, adiabatic compression of air from a temperature of 18 °C when the volumetric compression ratio is (a) 6, (b) 15.

Solution

Isentropic temperature ratio of compression, $\rho_v = r_v^{\gamma-1}$,

where r_v = volumetric compression ratio

(a) $\rho_v = r_v^{\gamma-1} = 6^{0.4} = 2.0477$

$(T_2 - T_1) = T_1 (\rho_v - 1) = 291.15 \times 1.0477 = \underline{305}$ K

(b) $\rho_v = 15^{0.4} = 2.9542$

$(T_2 - T_1) = 291.15 \times 1.9542 = \underline{569}$ K

4.2 The specific fuel consumption of a spark-ignition engine at full load is 0.093 kg per MJ of brake work output when the calorific value of the fuel is 44 MJ/kg. The mechanical efficiency is 80 %. Calculate the indicated overall efficiency. Also express the specific fuel consumption in (1) kg/kW h, (2) lb/hp h.

Calculate the b.m.e.p. and the i.m.e.p. of the engine at this load if the gravimetric air-fuel ratio is 18, the volumetric efficiency is 82 % and the ambient air conditions are 1 bar and 18 °C.

The engine has a volumetric compression ratio of 6. Determine the thermal efficiency of the comparable air-standard Otto cycle and thence the indicated efficiency ratio for the engine.

Calculate the i.m.e.p., maximum temperature and maximum pressure for the Otto cycle if the pressure and temperature at the beginning of compression are respectively 1 bar and 18 °C, and the heat supplied per unit mass of air is equal to the energy supplied in the engine by the fuel (in terms of its calorific value) per unit mass of air drawn in.

Solution

$$m_f \, CV \, \eta_o \, \eta_M = W_B,$$

where $m_f \equiv$ fuel consumption, kg

$CV \equiv$ calorific value, MJ/kg

$\eta_o \equiv$ brake overall efficiency

$\eta_M \equiv$ mechanical efficiency

$W_B \equiv$ brake work, MJ

$$\therefore \text{ Specific fuel consumption } \equiv \frac{m_f}{W_B} = \frac{1}{CV \, \eta_o \, \eta_M} \text{ kg/MJ}$$

$$\therefore \eta_o = \frac{100}{0.093 \times 44 \times 0.80} = \underline{30.5} \ \%$$

1 kW h = 3.6 MJ

(1) Specific fuel consumption = 0.093 x 3.6 = $\underline{0.335}$ kg/kW h

1 lb/hp h = 0.169 kg/MJ (Appendix D of Ref. 2)

(2) Specific fuel consumption $= \dfrac{0.093}{0.169} = \underline{0.550}$ lb/hp h

$$\text{b.m.e.p. } (N/m^2) \equiv \frac{\text{Brake work output } (J \equiv N\,m)}{\text{Swept volume } (m^3)}$$

Specific volume of air at 1 bar (10^5 N/m²) and 18 °C (291 K)

$$= \frac{RT}{p} = \frac{0.287 \times 10^3 \times 291}{10^5} = 0.8352 \text{ m}^3/\text{kg}$$

<u>Per MJ of brake work output:</u>

Air consumption = Specific fuel consumption x Air-fuel ratio

= 0.093 x 18 = 1.674 kg

= 1.674 x 0.8352 = 1.398 m³

Volumetric efficiency = 0.82 (given)

$$\therefore \text{ Suction swept volume} = \frac{\text{Air consumption } (m^3)}{\text{Volumetric efficiency}} = \frac{1.398}{0.82} = 1.705 \text{ m}^3$$

$$\therefore \text{ b.m.e.p.} = \frac{1}{1.705} = \underline{0.587} \text{ MN/m}^2$$

$$\text{i.m.e.p} = \frac{\text{b.m.e.p}}{\eta_M} = \frac{0.587}{0.8} = \underline{0.733}\ \text{MN/m}^2$$

Volumetric compression ratio $\equiv r_v$

<u>In the Otto cycle</u>, isentropic temperature ratio, $\rho_v = r_v^{\gamma-1}$

$$= 6^{0.4} = 2.0477$$

$$\eta_{OTTO} = (1 - \frac{1}{\rho_v}) = (1 - \frac{1}{2.0477}) \times 100 = \underline{51.2}\ \%$$

$$\text{Indicated efficiency ratio} \equiv \frac{\eta_o}{\eta_{OTTO}} = \frac{30.5}{51.2} = \underline{0.596}$$

<u>Per kg of air</u>, $Q_{in} = \dfrac{\text{CV}}{\text{Air/fuel ratio}} = \dfrac{44}{18}$ MJ

$$W_{net} = Q_{in} \times \eta_{OTTO} = \frac{44}{18} \times 0.512 = 1.251 \text{ MJ}$$

Clearance volume $\equiv v_0$ m³

Volumetric compression ratio, $r_v = 6$ (given)

Specific volume of air = 0.8352 m³/kg (above)

$$\therefore v_0 = \frac{0.8352}{6}$$

$$\therefore \text{Swept volume} = 5\,v_0 = \frac{5 \times 0.8352}{6} = 0.696 \text{ m}^3$$

$$\therefore \text{i.m.e.p.} \equiv \frac{W_{net}}{\text{Swept volume}} = \frac{1.251}{0.696} = \underline{1.80}\ \text{MN/m}^2$$

<u>Using the notation of Fig. 4.4(b)</u>:

$t_1 = 18$ °C, $(t_2 - t_1) = 305$ K $\quad \therefore t_2 = 323$ °C

<u>Per kg of air</u>, $Q_{in} = \dfrac{\text{Calorific value}}{\text{Air/fuel ratio}} = \dfrac{44 \times 10^3}{18}$ kJ/kg

$$c_v = 0.72 \text{ kJ/kg K}$$

$$(t_3 - t_2) = \frac{Q_{in}}{c_v} = \frac{44 \times 10^3}{18 \times 0.72} = 3395 \text{ K}$$

$$\therefore \text{Max. temperature, } t_3 = 3395 + 323 \approx \underline{3720}\ °\text{C}$$

$$p_3 = p_1 \times \frac{v_1}{v_3} \times \frac{T_3}{T_1}, \quad T_1 = 291 \text{ K},\ T_3 = 3993 \text{ K}$$

$$\text{Max. pressure, } p_3 = 1 \times 6 \times \frac{3993}{291} = 82.3 \text{ bar} = \underline{8.23}\ \text{MN/m}^2$$

4.3 The specific fuel consumption of a compression-ignition engine at full load is 0.068 kg per MJ of brake work output when the calorific value of the fuel is 44 MJ/kg. The mechanical efficiency is 80 %. Calculate the indicated overall efficiency. Also express the specific fuel consumption in (1) kg/kW h, (2) lb/hp h.

Calculate the b.m.e.p. and i.m.e.p. of the engine at this load if the gravimetric air-fuel ratio is 28, the volumetric efficiency is 82 %, and ambient air conditions are 1 bar and 18 °C.

The engine has a volumetric compression ratio of 15. Determine the thermal efficiency of the following air-standard cycles having the same volumetric compression ratio as the engine, and thence calculate the indicated efficiency ratio for the engine with respect to each of these cycles:

(a) an Otto cycle;
(b) a Diesel cycle in which the temperature at the beginning of compression is 18 °C, and in which the heat supplied per unit mass of air is equal to the energy supplied in the engine by the fuel (in terms of its calorific value) per unit mass of air drawn in.

Solution

The calculations are on similar lines to those presented in the Solution to Problem 4.2.

$$\eta_o = \frac{100}{0.068 \times 44 \times 0.80} = \underline{41.8}\ \%$$

(1) Specific fuel consumption $= 0.068 \times 3.6 = \underline{0.245}$ kg/kW h

(2) Specific fuel consumption $= \dfrac{0.068}{0.169} = \underline{0.402}$ lb/hp h

Per MJ of brake work output:

$$\text{Air consumption} = 0.068 \times 28 = 1.904 \text{ kg}$$

$$= 1.904 \times 0.8352 = 1.590 \text{ m}^3$$

$$\text{Suction swept volume} = \frac{1.590}{0.82} = 1.939 \text{ m}^3$$

$$\text{b.m.e.p.} = \frac{1}{1.939} = \underline{0.516} \text{ MN/m}^2$$

$$\text{i.m.e.p.} = \frac{0.516}{0.8} = \underline{0.645} \text{ MN/m}^2$$

(a) In the Otto cycle:

$$\text{Isentropic temperature ratio} \equiv \rho_v = r_v^{\gamma-1} = 15^{0.4} = 2.9542$$

$$\eta_{OTTO} = (1 - \frac{1}{\rho_v}) = (1 - \frac{1}{2.9542}) \times 100 = \underline{66.2}\ \%$$

$$\text{Indicated efficiency ratio} \equiv \frac{\eta_o}{\eta_{OTTO}} = \frac{41.8}{66.2} = \underline{0.632}$$

(b) In the Diesel cycle:

Using the notation of Fig. 4.5:

Process 1-2 (Compression): $T_1 = 291$ K

$$T_2 = \rho_v T_1 = 291 \times 2.954 = 860 \text{ K}$$

Process 2-3 (Heat supplied):

Per kg of air, $Q_{in} = \dfrac{44 \times 10^3}{28} = 1571.4$ kJ/kg

$$(T_3 - T_2) = \frac{Q_{in}}{c_p} = \frac{1571.4}{1.01} = 1556 \text{ K}$$

$$T_3 = 1556 + 860 = 2416 \text{ K}$$

$$v_3 = v_0 \times \frac{T_3}{T_2} = v_0 \times \frac{2416}{860} = 2.809\, v_0$$

Process 3-4 (Expansion):

$$T_4 = \left(\frac{v_3}{v_4}\right)^{\gamma-1} T_3 = \left(\frac{2.809\, v_0}{15\, v_0}\right)^{0.4} \times 2416$$

$$= \frac{2416}{(5.340)^{0.4}} = \frac{2416}{1.9544} = 1236 \text{ K}$$

$$(T_4 - T_1) = 1236 - 291 = 945 \text{ K}$$

Diesel cycle efficiency, $\eta_{DIESEL} = 1 - \dfrac{Q_{out}}{Q_{in}}$

$$= 1 - \frac{c_v(T_4 - T_1)}{c_p(T_3 - T_2)} \quad \text{But } \frac{c_p}{c_v} = \gamma = 1.4$$

$$\therefore \eta_{DIESEL} = \left(1 - \frac{945}{1.4 \times 1556}\right) \times 100 = \underline{56.6\ \%}$$

$$\text{Indicated efficiency ratio} \equiv \frac{\eta_o}{\eta_{DIESEL}} = \frac{41.8}{56.6} = \underline{0.739}$$

4.4 Calculate the i.m.e.p., maximum temperature and maximum pressure for each of the cycles of Problem 4.3 when the temperature at the beginning of compression and the heat supplied per unit mass of air are the same for both cycles, and the pressure at the beginning of compression is 1 bar.

Tabulate the values of thermal efficiency, maximum temperature, maximum pressure and i.m.e.p. for the air-standard cycles of Problems 4.2 and 4.3, and comment on the figures.

Solution

(a) For the Otto cycle of Problem 4.3:

Q_{in} = 1.5714 MJ/kg of air (As for the Diesel cycle of Problem 4.3)

$W_{net} = \eta_{OTTO}\, Q_{in} = 0.662 \times 1.5714 = 1.040$ MJ/kg

Using the notation of Fig. 4.4(b):

At start of compression (state-point 1), per kg of air:

$15\, v_0 = 0.8352$ (Problem 4.2)

$\therefore$ Swept volume $= \frac{14}{15} \times 0.8352 = 0.7795$ m³

$$\text{i.m.e.p} \equiv \frac{\text{Area of diagram}}{\text{Swept volume}} = \frac{W_{net}}{\text{Swept volume}} = \frac{1.040}{0.7795} = \underline{1.33}\ \text{MN/m}^2$$

$$T_3 - T_2 = \frac{Q_{in}}{c_v} = \frac{1571.4}{0.72} = 2183\ \text{K}$$

$T_2 = 860$ K (As for the Diesel cycle of Problem 4.3)

Max. temp., $T_3 = 860 + 2183 = 3043$ K $\quad \therefore t_3 = \underline{2770}$ °C

$$\text{Max. pressure, } p_3 = p_1 \times \frac{v_1}{v_3} \times \frac{T_3}{T_1} = 0.1 \times 15 \times \frac{3043}{291} = \underline{15.7}\ \text{MN/m}^2$$

(b) <u>For the Diesel cycle of Problem 4.3</u>:

Q_{in} = 1571.4 kJ/kg

$W_{net} = \eta_{DIESEL}\, Q_{in} = 0.566 \times 1.5714 = 0.8894$ MJ/kg

Using the notation of Fig. 4.5:

Swept volume = 0.7795 m³ (as for Otto cycle)

$$\text{i.m.e.p.} \equiv \frac{W_{net}}{\text{Swept volume}} = \frac{0.8894}{0.7795} = \underline{1.14}\ \text{MN/m}^2$$

Max. temperature, $T_3 = 2416$ K (Problem 4.3) $\therefore t_3 = \underline{2143}$ °C

$$\text{Max. pressure, } p_2 = p_1 \times \frac{v_1}{v_2} \times \frac{T_2}{T_1} = 0.1 \times 15 \times \frac{860}{291} = \underline{4.43}\ \text{MN/m}^2$$

Engine air-fuel ratio	Air-standard cycle	r_v	η_{CY} %	$t_{max.}$ °C	$p_{max.}$ MN/m²	i.m.e.p. MN/m²
18 (S.I.)	Otto	6	51.2	3720	8.23	1.80
28 (C.I.)	(Otto)	(15)	(66.2)	(2770)	(15.7)	(1.33)
	Diesel	15	56.6	2143	4.43	1.14

S.I. = spark-ignition C.I. = compression-ignition

4.5 A six-cylinder, four-stroke petrol engine is to develop 40 kW at 40 rev/s when designed for a volumetric compression ratio of 6.0. The ambient air conditions are 1 bar and 18 °C, and the calorific value of the fuel is 44 MJ/kg.

(a) Calculate the specific fuel consumption in kg per MJ of brake work output if the indicated overall efficiency is estimated to be 60 % of the thermal efficiency of the corresponding air-standard Otto cycle and the estimated mechanical efficiency is 80 %.

(b) The required gravimetric air-fuel ratio is 15.4 and the estimated volumetric efficiency is 82 %. Determine the required total swept volume, and the cylinder bore if the bore is to be equal to the stroke.

(c) Calculate the b.m.e.p.

Solution

(a) $\eta_{OTTO} = 0.512$ (Problem 4.2)

$\therefore$ Indicated overall efficiency, $\eta_o = 0.6 \times 0.512 = 0.3072$

As in Problem 4.2, $m_f \, CV \, \eta_o \, \eta_M = W_B$

$$\therefore \text{Specific fuel consumption} = \frac{m_f}{W_B} = \frac{1}{44 \times 0.3072 \times 0.80}$$

$$= \underline{0.0925} \text{ kg/MJ}$$

(b) Rate of air supply $= 15.4 \times 0.0925 \times 10^{-3} \times 40 = 0.0570$ kg/s

Specific volume of air at 1 bar and 18 °C $= 0.8352$ m³/kg (Problem 4.2)

Rate of air supply (m³/s) = swept volume (m³) $\times \frac{\text{rev/s}}{2} \times$ volumetric efficiency

$$\therefore \text{Swept volume} = \frac{0.0570 \times 0.8352}{20 \times 0.82} \times 10^6 = \underline{2903} \text{ cm}^3$$

If d = bore (cm), stroke = bore, and number of cylinders = 6:

$$\frac{\pi d^2}{4} \times d = \frac{2903}{6}, \qquad d^3 = 616.0$$

$$\therefore \text{Bore} = 8.51 \text{ cm} = \underline{85} \text{ mm}$$

(c) Brake power $= p_e LAN'$ [Equation (4.20)]

$$\therefore \text{b.m.e.p.} \equiv p_e = \frac{40 \times 10^3}{2903 \times 10^{-6} \times 20} \times 10^{-6} = \underline{0.689} \text{ MN/m}^2$$

4.6 The four-stroke petrol engine in an automobile has a total swept volume of V litres. The diameter of the road wheels is d m. At cruising speed the ratio of the engine speed to the speed of the road wheels is n, the volumetric efficiency of the engine is η_v and the carburettor maintains a gravimetric air-fuel ratio of f. The specific volume of the ambient air is v m³/kg and the specific gravity of the fuel is σ. Show that the "fuel consumption", in km/litre, is equal to

$$2\pi \frac{\sigma \, f \, d \, v}{n \, \eta_v \, V}.$$

Such an engine has a volumetric efficiency of 50 % when the cruising speed of the automobile is 65 km/h. The brake overall efficiency of the engine (based on a calorific value of the fuel of 45 MJ/kg) is then estimated to be 28 %. Calculate the "fuel consumption" and the power output of the engine under these conditions, given that $\sigma = 0.7$, f = 17, d = 0.7, v = 0.8, n = 4.6 and V = 2.1.

Solution

Per km travelled:

$$\text{Engine revolutions} = \frac{1000 \, n}{\pi d}$$

$$\text{Volume swept through} = \frac{1}{2} \times \frac{1000 \, n}{\pi d} \times V \times 10^{-3} = \frac{nV}{2\pi d} \text{ m}^3$$

$$\text{Air inhaled} = \frac{nV}{2\pi d} \times \frac{\eta_v}{v} \text{ kg}$$

$$\text{Fuel consumed} = \frac{1}{2\pi} \frac{n \, \eta_v \, V}{d \, v} \times \frac{1}{f} \text{ kg}$$

$$\text{Density of water} = 1 \text{ kg/litre}$$

$$\therefore \text{Density of fuel} = \sigma \text{ kg/litre}$$

$$\therefore \text{Fuel consumed} = \frac{1}{2\pi} \frac{n \, \eta_v \, V}{\sigma \, f \, d \, v} \text{ litre per km travelled}$$

$$\therefore \text{"Fuel consumption"} = 2\pi \frac{\sigma \, f \, d \, v}{n \, \eta_v \, V} \text{ km/litre}$$

Inserting the values given,

$$\text{"Fuel consumption"} = 2\pi \frac{0.7 \times 17 \times 0.7 \times 0.8}{4.6 \times 0.5 \times 2.1} = \underline{8.67} \text{ km/litre}$$

Since the density of the fuel = 0.7 kg/litre

$$\text{Rate of fuel consumption} = \frac{65}{3600} \times \frac{0.7}{8.67} = 1.458 \times 10^{-3} \text{ kg/s}$$

Given that calorific value of fuel = 45×10^3 kJ/kg, and brake η_o = 28 %.

$$\therefore \text{Power output} = 1.458 \times 10^{-3} \times 45 \times 10^3 \times 0.28 = \underline{18.4} \text{ kW}$$

4.7 The following particulars relate to a test on an open-circuit (internal combustion) gas-turbine plant, in which liquid n-octane (C_8H_{18}) of lower calorific value 44.43 MJ/kg was supplied to the adiabatic combustion chamber at a temperature of 25 °C:

Compressor:	Pressure ratio	4.13
	Air inlet temperature	290 K
	Air exit temperature	460 K
Turbine:	Gas inlet temperature	1000 K
	Gas exit temperature	750 K

Calculate:

(a) The isentropic efficiency of the compressor;
(b) The moles of air supplied to the combustion chamber by the compressor per mole of fuel burned, and thence the percentage of excess air;
(c) The turbine work output and compressor work input per kg of fuel burned;
(d) The overall efficiency of the plant, neglecting mechanical losses;
(e) The thermal efficiency of the corresponding air-standard Joule cycle, and thence the efficiency ratio for the plant.

The air passing through the compressor may be treated as a perfect gas for which $\gamma = 1.4$, $c_p = 1.01$ kJ/kg K and the molar mass is 29.0 kg/kmol. The enthalpies of the gases passing through the turbine, in MJ/kmol, are given in the following table:

Temperature K	O_2	N_2	CO_2	H_2O
1000	31.37	30.14	42.78	35.90
750	22.83	22.17	29.65	26.00
298	8.66	8.67	9.37	9.90

Solution

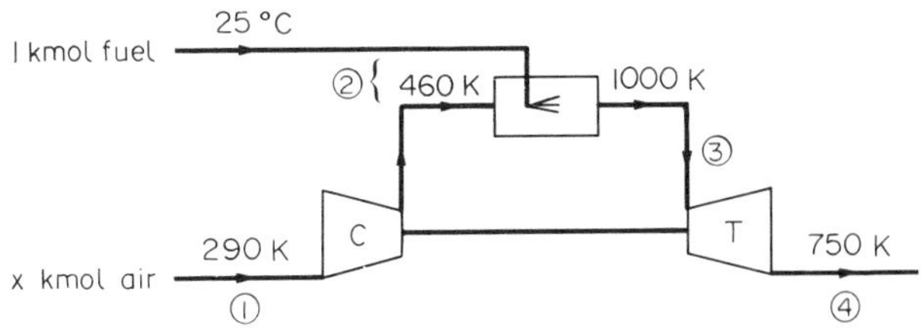

Fig. P.4.7(a)

(a) Compressor Pressure ratio, $r_p = 4.13$ (given)

Isentropic temperature ratio, $\rho_p = r_p^{(\gamma-1)/\gamma} = 4.13^{0.4/1.4} = 1.500$

$$(T_{2_s} - T_1) = T_1\ (\rho_p - 1) = 290 \times 0.5 = 145\ \text{K}$$

$$(T_2 - T_1) = (460 - 290) = 170\ \text{K}$$

$$\therefore \text{ Compressor efficiency, } \eta_c = \frac{c_p(T_{2_s} - T_1)}{c_p(T_2 - T_1)} = \frac{145}{170} \times 100 = \underline{85.3\ \%}$$

(b) <u>Combustion chamber</u>

$$C_8H_{18} + 12.5\ O_2 = 8\ CO_2 + 9\ H_2O$$

Let x kmol of air be supplied <u>per kmol of fuel</u>; then the kilomoles of products are:

Excess $O_2 = (0.21x - 12.5)$, $N_2^* = 0.79x$, $CO_2 = 8$, H_2O vapour $= 9$,

where the symbol N_2^* denotes 'atmospheric nitrogen' (see Ref. 2).

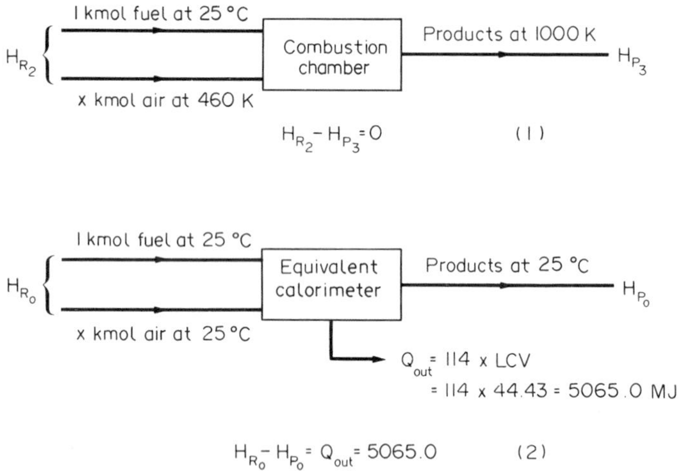

Fig. P.4.7(b)

Subtracting equation (1) from equation (2):

$$(H_{P_3} - H_{P_0}) - (H_{R_2} - H_{R_0}) = 5065.0 \qquad (3)$$

<u>Enthalpies of reactants</u>

Air: $(H_2 - H_0) = 29x \times 1.01 \times 10^{-3} \times (460 - 298) = 4.75x$

Fuel: $(H_2 - H_0) = 0$

$$\therefore (H_{R_2} - H_{R_0}) = \underline{4.75x}\ \text{MJ}$$

Enthalpies of products (using given data):

$$\begin{aligned}
\text{Excess } O_2:\ (H_3 - H_0) &= (0.21x - 12.5)(31.37 - 8.66) = 4.77x - 283.9 \text{ MJ}\\
N_2^*:\ (H_3 - H_0) &= 0.79x\ (30.14 - 8.67) = 16.96x\\
CO_2:\ (H_3 - H_0) &= 8\ (42.78 - 9.37) = 267.3\\
H_2O:\ (H_3 - H_0) &= 9\ (35.90 - 9.90) = 234.0\\
\therefore\ (H_{P_3} - H_{P_0}) &= 21.73x + 217.4 \text{ MJ}
\end{aligned}$$

Substituting these values in equation (3):

$$(21.73x + 217.4) - 4.75x = 5065.0$$

$$\therefore \text{ Moles of air per mole of fuel} \equiv x = \frac{4847.6}{16.98} = \underline{285.5}$$

$$\text{Theoretical (stoichiometric) air} = \frac{12.5}{0.21} = 59.5 \text{ kmol}$$

$$\therefore \text{ Excess air} = \frac{285.5 - 59.5}{59.5} \times 100 = \underline{380}\ \%$$

Combustion products are:

$$\begin{aligned}
\text{Excess } O_2 &= 0.21 \times 285.5 - 12.5 = 47.5 \text{ kmol}\\
N_2^* &= 0.79 \times 285.5 = 225.5\\
CO_2 &= 8\\
H_2O &= 9
\end{aligned}$$

(c) Turbine and compressor

Enthalpy drops in turbine:

$$\begin{aligned}
\text{Excess } O_2 &= 47.5(31.37 - 22.83) = 405.7 \text{ MJ}\\
N_2^* &= 225.5(30.14 - 22.17) = 1797.2\\
CO_2 &= 8\ (42.78 - 29.65) = 105.0\\
H_2O &= 9\ (35.90 - 26.00) = 89.1
\end{aligned}$$

Total enthalpy drop in turbine = 2397.0 MJ per kmol of fuel

$$\therefore \text{ Per kg of fuel, turbine work output} = \frac{2397.0}{114} = \underline{21.03} \text{ MJ}$$

Compressor

$$\text{Per kg of fuel, compressor work input} = \frac{(285.5 \times 29.0) \times 1.01 \times 170}{114} \times 10^{-3}$$

$$= \underline{12.47} \text{ MJ}$$

(d) Overall efficiency,

$$\eta_o \equiv \frac{W_{net}}{CV} = \frac{21.03 - 12.47}{44.43} \times 100 = \underline{19.3}\ \%$$

(e) Air-standard Joule cycle

Thermal efficiency, $\eta_{JOULE} = (1 - \frac{1}{\rho_p})$ [Equation 3.4 in Section 3.4]

$$= (1 - \frac{1}{1.5}) \times 100 = \underline{33.3}\ \%$$

$$\text{Efficiency ratio} \equiv \frac{\eta_o}{\eta_{JOULE}} = \frac{19.3}{33.3} = \underline{0.580}$$

4.8 An isothermal, reversible fuel cell takes in hydrogen and oxygen, each at 1 atm and 25 °C, and delivers water at 1 atm and 25 °C. For this reaction, $-\Delta G_0$ = 117.6 and $-\Delta H_0$ = 142.0 MJ per kg of hydrogen consumed. Calculate the heat transfer from the cell to its environment at 25 °C, per kg of hydrogen consumed.

A hydrogen-oxygen fuel cell operating in an environment at 25 °C consumes hydrogen at a rate of 0.36 litre/min when delivering a current of 50 amperes at a p.d. of 0.8 V. The specific volume of the hydrogen is 12.14 m³/kg. Calculate (a) the rational efficiency, (b) the arbitrary overall efficiency of the fuel cell, based on the higher calorific value of hydrogen.

Solution

$W_{REV} = -\Delta G_0$ [Equation (4.1)]

If $H_0 \equiv$ enthalpy of reactants (hydrogen and oxygen)

$H_0 + \Delta H_0$ = enthalpy of product (water)

Then $Q_{out} = -\Delta H_0 - W_{REV}$

$\therefore Q_{out} = -\Delta H_0 + \Delta G_0 = 142.0 - 117.6 = \underline{24.4}$ MJ per kg of hydrogen

For the given fuel cell:

(a) Ideal work output, $W_{REV} = -\Delta G_0$ = 117.6 MJ per kg of hydrogen

$$\text{Mass flow rate of hydrogen} = \frac{0.36 \times 10^{-3}}{60 \times 12.14} = 0.4942 \times 10^{-6}\ \text{kg/s}$$

$\therefore$ Ideal power output $= 0.4942 \times 10^{-6} \times 117.6 \times 10^{6} = 58.12$ W

Actual power output $= 50 \times 0.8 = 40.0$ W

$$\therefore \text{Rational efficiency, } \eta_R = \frac{40.0}{58.12} \times 100 = \underline{68.8}\ \%$$

(b) Arbitrary overall efficiency, $\eta_o \equiv \frac{W_{actual}}{-\Delta H_0}$

$$= \frac{-\Delta G_0}{-\Delta H_0} \times \frac{W_{actual}}{-\Delta G_0}$$

$$= \frac{-\Delta G_0}{-\Delta H_0}\,\eta_R = \frac{117.6}{142.0} \times 68.8 = \underline{57.0}\ \%$$

4.9 In a steady-flow, reversible fuel cell operating isothermally at 25 °C and atmospheric pressure, hydrogen and oxygen gas streams enter and water leaves.

Calculate the e.m.f. of the cell at 25 °C, given that the electronic charge is 1.60×10^{-19} coulomb and that the molar number (Avogadro constant) is 6.02×10^{26} per kmol.

Determine the required rate of supply of hydrogen, in litres per minute at 1 atm pressure and 25 °C, for such a cell of 100 W output, and calculate the rate of heat transfer with the environment, in watts.

Solution

For each molecule of hydrogen consumed, 2 electrons pass across the cell from the positive electrode to the negative electrode (see Ref. 3, page 385).

∴ Per kmol of hydrogen consumed:

$$\text{Charge transferred} = 2 \times 1.60 \times 10^{-19} \times 6.023 \times 10^{26}$$

$$= 1.927 \times 10^{8} \text{ coulomb}$$

$$\text{Mass transferred} = 2 \text{ kg}$$

Per kg of hydrogen, $W_{REV} = -\Delta G_0 = 117.6$ MJ

Then, if the e.m.f. of the cell is E volts:

$$1.927 \times 10^{8} \times E = 2 \times 117.6 \times 10^{6}$$

$$\therefore \text{ e.m.f. of cell} = \frac{2.352}{1.927} = \underline{1.22}\ \text{V}$$

For 100 W output:

$$\text{Flow rate of hydrogen} = \frac{100}{W_{REV}} = \frac{100}{-\Delta G_0} = \frac{100}{117.6 \times 10^{6}} \text{ kg/s}$$

$$\text{Specific volume of hydrogen} = 12.14 \text{ m}^3/\text{kg}$$

$$\therefore \text{ Flow rate of hydrogen} = \frac{100}{117.6 \times 10^{6}} \times 12.14 \times 10^{3} \times 60 = \underline{0.62} \text{ litre/min}$$

If $\dot{m} \equiv$ flow rate of hydrogen,

$\dot{W}_{REV} \equiv$ power output of cell $= \dot{m}(-\Delta G_0)$,

$\dot{Q}_{out} \equiv$ rate of heat transfer from cell to environment,

then $\dot{Q}_{out} = (\dot{H}_R - \dot{H}_P) - \dot{W}_{REV}$, where $\dot{H}_R - \dot{H}_P = \dot{m}(-\Delta H_0)$

$$\therefore \dot{Q}_{out} = \dot{W}_{REV}\left(\frac{\dot{H}_R - \dot{H}_P}{\dot{W}_{REV}} - 1\right) = \dot{W}_{REV}\left(\frac{-\Delta H_0}{-\Delta G_0} - 1\right)$$

$\therefore$ Rate of heat transfer to environment $= 100\left(\frac{142.0}{117.6} - 1\right) = \underline{20.7}$ W

CHAPTER 5

Simple Refrigerating Plant

5.1 Show that the coefficient of performance of a refrigerating plant which operates between two thermal-energy reservoirs, each at uniform temperature, cannot be greater than that of a reversible plant operating between the same two reservoirs.

Solution

Consider two cyclic refrigerating plants operating between the same two reservoirs A and B, where $T_A < T_B$, plant R being reversible and I irreversible.

Let them both take in heat Q_A from reservoir A.

Let R take in net work W_R and I take in net work W_I,

and R and I deliver heat quantities $(Q_B)_R$ and $(Q_B)_I$ respectively to reservoir B.

$$\text{Coefficient of performance, CP} \equiv \frac{Q_{in}}{W_{in}}$$

Hence, for the same Q_A taken in by R and I:

Suppose that $CP_I > CP_R$

then $W_R > W_I$

and $(Q_B)_R > (Q_B)_I$.

Since R is reversible, all the effects on the plant and the environment can be effaced. Let E be the "effacer", namely the cyclic heat power plant which would efface all the effects of R (i.e. E would take in heat $(Q_B)_R$ from reservoir B, deliver work W_R and reject heat Q_A to reservoir A).

Reservoir A is now superfluous, since E delivers to it the heat Q_A

drawn from it by I. Thus E and I together take a positive net quantity of heat $[(Q_B)_R - (Q_B)_I]$ from reservoir B and deliver a positive net quantity of work $(W_R - W_I)$, while not rejecting any heat to a second reservoir. They together thus constitute a cyclic perpetual motion machine of the second kind, a cyclic PMM 2. That would constitute a contravention of the Second Law of Thermodynamics.

Hence, contrary to the supposition, CP_I cannot be greater than CP_R.

5.2 Repeat the calculations of Example 5.1 when the refrigerant is (1) ammonia, (2) methyl chloride and (3) Refrigerant-12.

Solution

Example 5.1 related to a refrigerating plant operating on the quasi-ideal vapour-compression cycle defined in Section 5.9, and in which the saturation temperatures in the evaporator and condenser were respectively -20 °C and 25 °C. The volumetric efficiency of the compressor was stated to be 100 %.

Using the notation of Fig. 5.4, the required calculations are set out below:

REFRIGERANT		Ammonia	Methyl Chloride	Refrigerant-12
h_1	kJ/kg	1420	455.2	178.7
$h_4 = h_3$	kJ/kg	299	100.5	59.7
(a) Refrig. effect/unit mass = $(h_1 - h_4)$	kJ/kg	*1121*	*354.7*	*119.0*
v_1	litre/kg	624	338	108.9
(b) Refrig. effect/unit vol.	kJ/litre	*1.80*	*1.05*	*1.09*
(c) Flow rate per tonne of refrigeration* = $3.86 \times 60/(h_1 - h_4)$ kg/min		*0.2066*	*0.6530*	*1.946*
(d) Compressor displacement per tonne of refrigeration = (c) x v_1 litre/min		*128.9*	*220.7*	*211.9*
$s_{2_s} = s_1$	kJ/kg K	5.624	1.803	0.7088
s' ≡ s at p_B and superheats of 50 K, 30 K and 0 K respectively		5.468	1.735	0.6869
s" ≡ s at p_B and superheats of 100 K, 60 K and 20 K respect.		5.800	1.820	0.7334
(s" - s')	kJ/kg K	0.332	0.085	0.0465
$(s_{2_s} - s')$	kJ/kg K	0.156	0.068	0.0219
(e) Superheat at 2_s (comp. exit)	K	*73*	*54*	*9*

*See the Solution to Problem 5.4.

REFRIGERANT		Ammonia	Methyl Chloride	Refrigerant-12
h_{2_s} (At p_B and the above superheat)	kJ/kg	1662	529.1	204.5
(f) W_{in} per unit mass $= (h_{2_s} - h_1)$	kJ/kg	*242*	*73.9*	*25.8*
(g) Power input/tonne $= \frac{(c) \times (f)}{60}$	kW	*0.833*	*0.804*	*0.837*
(h) CP = (a)/(f)		*4.63*	*4.80*	*4.61*

5.3 Repeat the calculations of Example 5.2 with the same refrigerants as in Problem 5.2.

Solution

Example 5.2 related to a refrigerating plant in which the throttle valve in the quasi-ideal vapour-compression cycle of Example 5.1 was replaced by an expander of 100 % isentropic efficiency. Thus the plant of Example 5.2 operated on the ideal vapour-compression cycle defined in Section 5.5.

Using the notation of Fig. 5.2 and data from Problem 5.2, the required calculations are set out below:

REFRIGERANT		Ammonia	Methyl Chloride	Refrigerant-12
$p_{cond.}$ at 25 °C	MN/m²	1.001	0.567	0.651
$p_{crit.}$	MN/m²	11.30	6.68	4.115
(1) $p_{cond.}/p_{crit.}$		*0.089*	*0.085*	*0.158*
$s_{4_s} = s_3 \equiv s_f$ at 25 °C	kJ/kg K	1.124	0.379	0.2239
s_f at -20 °C		0.368	0.124	0.0731
$(s_{4_s} - s_f)$		0.756	0.255	0.1508
s_g at -20 °C		5.624	1.803	0.7088
$(s_g - s_f)$ at -20 °C		5.256	1.679	0.6357
(2) Dryness fraction, $x_{4_s} = (s_{4_s}-s_f)/(s_g-s_f)$		*0.1438*	*0.1519*	*0.2372*

REFRIGERANT		Ammonia	Methyl Chloride	Refrigerant-12
h_{fg} at -20 °C	kJ/kg	1330.2	425.1	160.9
$x_{4_s} h_{fg}$		191.3	64.6	38.2
h_f at -20 °C		89.8	30.1	17.8
$h_{4_s} = h_f + x_{4_s} h_{fg}$		281.1	94.7	56.0
h_g at -20 °C		1420	455.2	178.7
Refrig. effect/unit mass $= (h_g - h_{4_s})$		1139	360.5	122.7
(3) Increase in refrigerating effect	%	*1.6*	*1.6*	*3.1*
$h_3 = h_f$ at 25 °C	kJ/kg	298.9	100.5	59.7
Expander $W_{out} = (h_3 - h_{4_s})$ = Decrease in net W_{in}	kJ/kg	17.8	5.8	3.7
(4) Decrease in net W_{in}	%	*7.4*	*7.8*	*14.3*
Net W_{in}/unit mass	kJ/kg	224	68.1	22.1
(5) CP ≡ (Refrig.effect/Net W_{in})		*5.08*	*5.29*	*5.55*
$(h_{2_s} - h_3)$	kJ/kg	1363	428.6	144.8
$(s_{2_s} - s_3)$	kJ/kg K	4.500	1.424	0.4849
(6) Mean temp. of heat rejection in condenser $= (h_{2_s} - h_3)/(s_{2_s} - s_3)$	K	*302.9*	*301.0*	*298.6*
(7) CP of reversed Carnot between -20 °C and 25 °C			*5.62* (= 253/45)	

5.4 Derive the relation between the power input per tonne of refrigeration (in kW/t) and the coefficient of performance of a refrigerating plant. The enthalpy of fusion of ice at 0 °C is 333.5 kJ/kg.

In a test on a vapour-compression refrigerating plant using methyl chloride as the refrigerant, the saturation temperatures in the evaporator and condenser are respectively -5 °C and 40 °C, and the power input per tonne of refrigeration is 0.93 kW/t. Determine the coefficient of performance, and compare it with that of the corresponding quasi-ideal cycle.

Assuming that irreversibilities in the actual plant occur only in the flow through the throttle valve and the compressor, calculate the isentropic efficiency of the compressor.

Solution

By definition, a tonne of refrigeration is the rate of thermal energy extraction corresponding to the production, in a period of 24 hours, of 1 tonne (1000 kg) of ice at 0 °C from water at the same temperature.

$$\therefore\ 1 \text{ tonne of refrigeration} = \frac{1000 \times 333.5}{24 \times 3600} = 3.86 \text{ kW}$$

i.e. Rate of thermal energy extraction, $\dot{Q}_{out} = 3.86$ kW/t

But $CP \equiv \dfrac{\dot{Q}_{out}}{\dot{W}_{in}}$ (Note: $\dot{Q}_{out} \equiv \dot{Q}_{in}$ to refrigerant)

$\therefore$ *CP x (Power input in kW/t) = 3.86*

In the given plant, $\dot{W}_{in} = 0.93$ kW/t

$$\therefore\ CP = \frac{3.86}{0.93} = \underline{4.15}$$

Quasi-ideal cycle

Using the notation of Fig. 5.4:

$s_{2_s} = s_1 = 1.742$ kJ/kg K

$$\text{Superheat at } 2_s = 30 + \frac{(1.742 - 1.696)}{(1.783 - 1.696)} \times 30 = 30(1 + \frac{0.046}{0.087})$$

$$= 45.9 \text{ K}$$

$$h_{2_s} = 513.5 + \frac{15.9}{30} \times 31.0 = 529.9 \text{ kJ/kg}$$

$$h_1 = 463.3 \text{ kJ/kg}$$

$$W_{in} = h_{2_s} - h_1 = 66.6 \text{ kJ/kg}$$

$h_4 = h_3 = 124.8$ kJ/kg $\quad Q_A = (h_1 - h_4) = 338.5$ kJ/kg

$$\text{Coefficient of performance, } CP \equiv \frac{Q_A}{W_{in}} = \frac{338.5}{66.6} = \underline{5.08}$$

$$\text{Isentropic efficiency of compressor, } \eta_c = \frac{\text{Actual CP}}{\text{Quasi-ideal CP}}$$

$$= \frac{4.15}{5.08} \times 100 = \underline{81.7}\ \%$$

5.5 What would be the coefficient of performance of a plant in which conditions were the same as for the quasi-ideal cycle of Problem 5.4, with the exception that the liquid leaving the condenser were under-cooled by 5 K?

Solution

Again using the notation of Fig. 5.4:

$$h_1 = 463.3 \text{ kJ/kg (Problem 5.4)}$$

$$h_{3'} \approx h_f \text{ at } 35\ °C = 116.7 \text{ kJ/kg}$$

$$Q'_A = (h_1 - h_{3'}) = 346.6 \text{ kJ/kg}$$

$$W_{in} = 66.6 \text{ kJ/kg (As in Problem 5.4)}$$

$$\therefore \text{ Coefficient of performance, CP} \equiv \frac{Q'_A}{W_{in}} = \frac{346.6}{66.6} = \underline{5.20}$$

5.6 A vapour-compression plant with Refrigerant-12 as the refrigerant is used as a heat pump to supply 30 kW to a building which is maintained at a mean temperature of 20 °C when the mean temperature of the outside air is 0 °C. There is a temperature difference of 5 K between the mean temperature of the outside air and the saturation temperature of the refrigerant in the evaporator, and also between the saturation temperature of the refrigerant in the condenser and the mean temperature of the building.

Saturated liquid enters the throttle valve and saturated vapour enters the compressor, which has an isentropic efficiency of 82 %. Calculate the coefficient of performance of the heat pump, and the power input to the compressor.

The compressor is driven by an electric motor, the combined efficiency of the motor and drive being 75 %. Express the electrical power input as a fraction of the power that would be required if the same energy input to the room were supplied by direct electrical heating.

Solution

In the required quasi-ideal cycle:

Condenser temperature = 25 °C Evaporator temperature = -5 °C

Using the notation of Fig. 5.4:

$$s_{2_s} = s_1 = 0.6991 \text{ kJ/kg K}$$

$$\text{Superheat at } 2_s = 0 + \frac{(0.6991 - 0.6869)}{(0.7334 - 0.6869)} \times 20 = \frac{0.0122}{0.0465} \times 20$$

$$= 5.25 \text{ K}$$

$$h_{2_s} = 197.7 + \frac{5.25}{20} \times 14.4 = 201.5 \text{ kJ/kg}$$

$$h_1 = 185.4 \text{ kJ/kg}$$

$$h_{2_s} - h_1 = 16.1 \text{ kJ/kg}$$

$$W_{in} = (h_2 - h_1) = \frac{h_{2_s} - h_1}{\eta_c} = \frac{16.1}{0.82} = 19.6 \text{ kJ/kg}$$

$$h_2 = 205.0 \text{ kJ/kg}$$

$$h_3 = 59.7 \text{ kJ/kg}$$

$$Q_B = (h_2 - h_3) = 145.3 \text{ kJ/kg}$$

$$\text{Coefficient of performance, CP} \equiv \frac{Q_B}{W_{in}} = \frac{145.3}{19.6} = \underline{7.4}$$

$$\text{Power input to compressor} = \frac{30}{7.4} = \underline{4.0} \text{ kW}$$

Per kg of Refrigerant-12:

$$\text{Electrical work input} = \frac{19.6}{0.75} = 26.1 \text{ kJ/kg}$$

$$\therefore \frac{\text{Electrical power input to motor}}{\text{Electrical power input to room}} = \frac{26.1}{145.3} = \underline{0.18}$$

5.7 In the open-circuit refrigerating plant of Problem 3.5 of Chapter 3, what advantage does this method of aircraft air-conditioning have over one in which the air is throttled at supercharger outlet and supplied direct to an air cooler and thence to the cabin.

Solution

The physical size of an air-conditioning plant is of prime importance in an aircraft. The bulkiest items in air-conditioning and refrigerating plants are usually the heat exchangers, because of the relatively small temperature differences across the tubes of the heat exchangers. The plant of Problem 3.5 enables this temperature difference to be increased appreciably, since the air enters the air cooler at t_2 = 94 °C after passing through the compressor. This compares favourably with the temperature of 55 °C at which the air would have entered the heat exchanger had it been passed from the supercharger outlet direct to the air cooler.

5.8 Figure 5.6 gives the flow diagram of a heat-pump installation for the air-conditioning of an underground vault housed in a disused quarry, and used during war-time for the storage of valuable works of art.

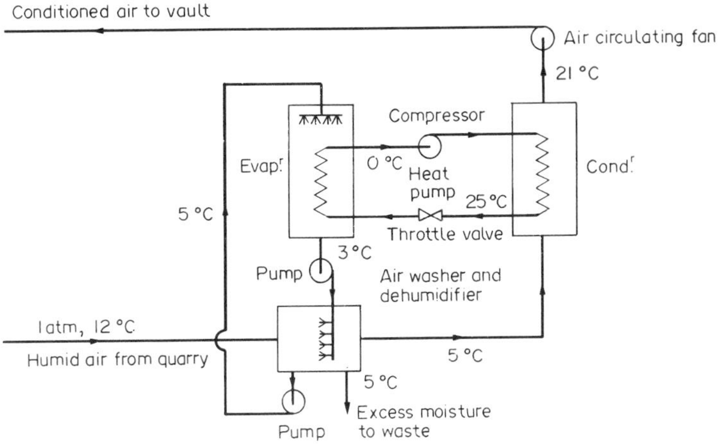

Fig. P.5.8 (Fig. 5.6)

Air of relative humidity 90 % and temperature 12 °C is drawn into the plant at a pressure of 1 atm through an air-washer and dehumidifier, where it is cooled by a water spray. The air leaves the washer saturated with water vapour at 5 °C and is then heated to 21 °C by being passed over the condenser coil of the heat pump, before being delivered to the vault.

The water is drawn from the air-washer at 5 °C and sprayed over the evaporator coil of the heat pump, where its temperature is reduced to 3 °C before return to the washer.

Calculate, per kg of air circulated:

(a) the amount of moisture extracted from the air in the washer;
(b) the amount of spray water;
(c) the coefficient of performance of the heat-pump installation;
(d) the ratio of the actual coefficient of performance to the theoretical coefficient of performance for a quasi-ideal cycle in which dry saturated Refrigerant-12 leaves the evaporator at 0 °C and the Refrigerant-12 is just condensed in the condenser at 25 °C.

Explain why a reduction in the specific humidity of the air is obtained by spraying water into it in the washer.

Solution

For the theoretical treatment (termed 'psychrometry') of air-vapour mixtures, Sections 16.17 to 16.20 in Chapter 16 of Ref. 3 may be consulted.

$$\text{Relative humidity, } \phi \equiv \frac{\text{Partial pressure of vapour in mixture}}{\text{Saturation pressure at mixture temperature}}$$

$$\equiv \frac{p'_v}{p_{sat.}}$$

(a) At washer inlet:

$\phi = 0.9$ (given), $p_{sat.}$ at 12 °C = 1.401 kN/m²

$\therefore\ p'_v = 0.9 \times 1.401 = 1.261$

Mixture pressure, $p = 101.325$ (1 atm)

$\therefore$ Partial pressure of air, $p'_a = 100.064$

Per kg of dry air:

$$\text{Mass of water vapour, } m_i = \frac{p'_v}{p'_a} \cdot \frac{M_v}{M_a}, \text{ where } M \equiv \text{molar mass (kg/kmol)}$$

$$= \frac{1.261}{100.06} \times \frac{18}{29} = 0.00782 \text{ kg}$$

At washer outlet:

$\phi = 1.0$ (given), $p_{sat.}$ at 5 °C = 0.874 kN/m²

$\therefore\ p'_v = 1.0 \times 0.874 = 0.874$ kN/m²

$$p = 101.325 \text{ kN/m}^2 \qquad (1 \text{ atm})$$

$$\therefore \; p_a' = 100.451 \text{ kN/m}^2$$

Per kg of dry air:

Mass of water vapour, $m_o = \dfrac{0.874}{100.45} \times \dfrac{18}{29} = 0.00540$ kg

$\therefore$ Moisture extracted in washer $= m_i - m_o = (0.00782 - 0.00540)$

$= \underline{0.00242}$ kg

(b) Let the amount of spray water $\equiv$ S kg

Then the steady-flow energy equation for the washer gives:

$$S(21.0 - 12.6) = 1.01 \times 7 + 0.00540\,(2523.6 - 2510.8) + 0.00242\,(2523.6 - 21.0)$$

$$8.4S = 7.070 + 0.069 + 6.056$$

$\therefore$ Amount of spray water, $S = \dfrac{13.195}{8.4} = \underline{1.57}$ kg

(c) Evaporator-cooler

The steady-flow energy equation for the evaporator-cooler gives:

Heat extracted from the water, $Q_A = 8.4S = 13.195$ kJ

Condenser-heater

The steady-flow energy equation for the condenser-heater gives:

Heat supplied to the air-vapour mixture,

$$Q_B = 1.01 \times 16 + 0.00540\,(2540.0 - 2510.8)$$

$$= 16.318 \text{ kJ}$$

Compressor

Compressor work input, $W_{in} = Q_B - Q_A = 3.123$ kJ

Heat pump

Coefficient of performance, $CP \equiv \dfrac{Q_B}{W_{in}} = \dfrac{16.318}{3.123} = \underline{5.23}$

(d) Quasi-ideal cycle (using Refrigerant-12)

Condenser temperature = 25 °C   Evaporator temperature = 0 °C

Using the notation of Fig. 5.4:

$$s_{2_s} = s_1 = 0.6966 \text{ kJ/kg K}$$

$$\text{Superheat at } 2_s = 0 + \frac{(0.6966 - 0.6869)}{(0.7334 - 0.6869)} \times 20 = \frac{0.0097}{0.0465} \times 20 = 4.17 \text{ K}$$

$$h_{2_s} = 197.7 + \frac{4.17}{20} \times 14.4 = 200.7 \text{ kJ/kg}$$

$$h_1 = 187.5 \text{ kJ/kg}$$

$$W_{in} = (h_{2_s} - h_1) = 13.2 \text{ kJ/kg}$$

$$h_3 = 59.7 \text{ kJ/kg} \qquad Q_B = (h_{2_s} - h_3) = 141.0 \text{ kJ/kg}$$

$$\text{Theoretical CP} = \frac{141.0}{13.2} = 10.68$$

$$\underline{\text{Ratio}}: \frac{\text{Actual CP}}{\text{Theoretical CP}} = \frac{5.23}{10.68} = \underline{0.490}$$

By definition, the specific humidity, ω, of an air-vapour mixture is given by

$$\omega \equiv \frac{m_v}{m_a},$$

where m_v and m_a are respectively the masses of water vapour and of air in a given quantity of the mixture.

From the calculations in (a) above, it will be seen that ω is less at washer outlet than at washer inlet, since $m_o < m_i$. Thus we have the apparently paradoxical situation that, by spraying water into the mixture as it passes through the washer, the specific humidity of the mixture has been decreased. However, it can also be seen from (a) that this is because the water sprayed into the washer is colder than the entering air-vapour mixture, causing the mixture to be cooled from 12 °C to 5 °C. At 5 °C the saturation pressure of water substance is less than at 12 °C and the calculation in (a) shows that this is why the vapour content of the mixture is reduced in passing through the washer.

It is of interest to calculate the relative humidity of the conditioned air entering the vault.

In the circumstances under consideration, both air and water vapour may be treated as being ideal gases. The partial pressure of a particular component in a mixture of ideal gases is proportional to the mole fraction of that component in the mixture. Between washer outlet and vault inlet, there is no change in the mole fraction of either the air or the water vapour in the mixture, since no water vapour is removed between these two points. Thus:

$$p'_v \text{ at vault inlet} = p'_v \text{ at washer outlet} = 0.874 \text{ kN/m}^2$$

$$\text{Also, at vault inlet, } p_{sat.} = 2.49 \text{ kN/m}^2 \text{ (at 21 °C)}$$

$$\text{Hence, } \underline{\text{at vault inlet, relative humidity}}, \ \phi \equiv \frac{p'_v}{p_{sat.}} = \frac{0.874}{2.49} \times 100 = \underline{35\ \%}$$

Thus, the relative humidity of the air-vapour mixture has been reduced to a value of 35 % at inlet to the vault, compared to a value of 90 % in the quarry.

It is also of interest to record the fact that the plant illustrated in Fig. P.5.8 (Fig. 5.6) is that which was installed in a quarry in North Wales which housed the art treasures of the National Gallery in London during the Second World War of 1939 - 1945.

PART II

Advanced Power and Refrigerating Plants

CHAPTER 6

Advanced Gas-Turbine Plant

6.1 A CBTX cycle is designed for given values of T_a, T_b, η_C and η_T. The heat exchanger has an effectiveness of unity and causes negligible pressure drop. The plant is designed for the same pressure ratio as that for which a CBT cycle with the same values of the foregoing parameters would produce maximum work output per unit unit mass flow of circulating fluid, which may be treated as a perfect gas. Show that this CBTX cycle has the same efficiency as an ideal reversible Joule cycle designed for the same pressure ratio.

Solution

With reference to Fig. P.6.1 overleaf, since the effectiveness of the heat exchanger X is unity:

$$T_3 = T_5 \qquad \text{and} \qquad T_6 = T_2$$

$$\therefore\ Q_B = c_p\ (T_4 - T_5) \qquad \therefore\ Q_A = c_p\ (T_2 - T_1)$$

Turbine:

$$(T_4 - T_5) = \eta_T\ T_b(1 - \frac{1}{\rho_p})$$

Compressor:

$$(T_2 - T_1) = \frac{c_p\ T_a}{\eta_C}\ (\rho_p - 1)$$

where $\rho_p \equiv$ isentropic temperature ratio

Thermal efficiency of cycle

$$(1 - \eta_{CY}) = \frac{Q_A}{Q_B} = \frac{T_2 - T_1}{T_4 - T_5} = \frac{\rho_p}{\eta_C\ \eta_T\ \theta}, \text{ where } \theta \equiv T_b/T_a$$

But in this plant, it is given that ρ_p is that for which a CBT cycle with the same pressure ratio would produce maximum work output. Thus, from equation (3.14) in Section 3.7 of Ref. 1 [i.e. eqn. (4) in the Solution to Problem 3.4],

$\rho_p = \sqrt{\alpha}$, where $\alpha \equiv \eta_C \eta_T \theta$

$$(1 - \eta_{CY}) = \frac{\sqrt{\alpha}}{\alpha} = \frac{1}{\sqrt{\alpha}} \qquad (1)$$

For an ideal reversible Joule cycle,

$$\eta_{JOULE} = (1 - \frac{1}{\rho_p}) \qquad \text{[Equation (3.4)]}$$

∴ For a Joule cycle in which $\rho_p = \sqrt{\alpha}$,

$$(1 - \eta_{JOULE}) = \frac{1}{\sqrt{\alpha}} \qquad (2)$$

Comparing equations (1) and (2), the proposition is seen to be proved.

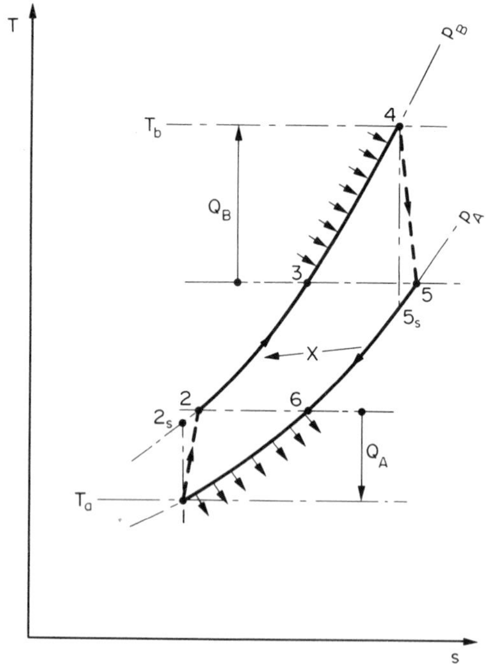

Fig. P.6.1.

6.2 The compressor and turbine inlet temperatures in a $(CBTX)_r$ cycle are respectively T_a and T_b. By sketching a temperature-entropy diagram, show that as $\rho_p \to 1$, $\eta_{CY} \to [1 - (1/\theta)]$, where $\theta \equiv T_b/T_a$.

Solution

In the $(CBTX)_r$ cycle, the effectiveness of the heat exchanger is unity.

$\therefore T_3 = T_5$ and $T_6 = T_2$

$\therefore Q_B = c_p (T_4 - T_5)$ $\therefore Q_A = c_p (T_2 - T_1)$

Turbine

$(T_4 - T_5) = T_b(1 - 1/\rho_p)$

Compressor

$(T_2 - T_1) = T_a(\rho_p - 1)$

where $\rho_p \equiv$ isentropic temperature ratio

Thermal efficiency of cycle

$$(1 - \eta_{CY}) = \frac{Q_A}{Q_B} = \frac{T_2 - T_1}{T_4 - T_5} = \frac{(\rho_p - 1)}{(1 - 1/\rho_p)} \frac{T_a}{T_b} = \frac{\rho_p}{\theta},$$

where $\theta \equiv T_b/T_a$

$$\therefore \eta_{CY} = [1 - (\rho_p/\theta)] \to [1 - (1/\theta)] \text{ as } \rho_p \to 1$$

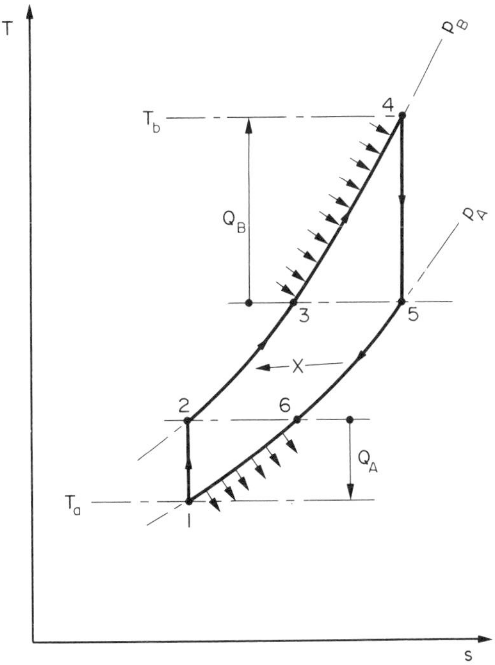

Fig. P.6.2.

6.3 Prove equation (6.5).

Solution

Equation (6.5) is: $\left(\frac{\eta_{CY}}{\eta_N} - 1\right) = \frac{[(\eta_R/\eta_N) - 1]}{[(Q_N/Q_R) + 1]}$,

where subscripts N and R relate to the hypothetical cycles in Fig. 6.8 for the $(CBTRT)_r$ cycle.

Using the notation of Fig. 6.8:

$$(1 - \eta_N) = (Q_{out})_{4'-1}/Q_N$$

$$(1 - \eta_R) = (Q_{out})_{6-4'}/Q_R$$

$$(1 - \eta_{CY}) = (Q_{out})_{6-1}/(Q_N + Q_R)$$

But $(Q_{out})_{6-1} = (Q_{out})_{6-4'} + (Q_{out})_{4'-1}$

Hence $(1 - \eta_{CY})(Q_N + Q_R) = (1 - \eta_R)Q_R + (1 - \eta_N)Q_N$

$$\therefore \eta_{CY} = \frac{\eta_R Q_R + \eta_N Q_N}{Q_N + Q_R}$$

$$\therefore (\eta_{CY} - \eta_N) = \frac{Q_R}{Q_N + Q_R}(\eta_R - \eta_N)$$

$$\therefore \left(\frac{\eta_{CY}}{\eta_N} - 1\right) = \frac{[(\eta_R/\eta_N) - 1]}{[(Q_N/Q_R) + 1]} \qquad (6.5)$$

6.4 By studying a sketch of a temperature-entropy diagram, show that the addition of a single stage of intercooling to a $(CBT)_r$ cycle will decrease the cycle efficiency.

Solution

The addition of a single stage of intercooling to a $(CBT)_r$ cycle produces a $(CICBT)_r$ cycle, which is discussed in Section 6.8 alongside the $(CBTRT)_r$ cycle illustrated in Fig. 6.8 and treated in Problem 6.3. In the latter case, the cycle was seen to be equivalent to a hypothetical simple $(CBT)_r$ cycle N to which had been added another hypothetical simple $(CBT)_r$ cycle R. The latter cycle has a lower pressure ratio than cycle N and so $\eta_R < \eta_N$, as will be seen from equation (3.4) in Section 3.4. It follows from equation (6.5) derived in Problem 6.3 that $\eta_{CY} < \eta_N$, so that the addition of reheating to a simple $(CBT)_r$ cycle does not improve the thermal efficiency. This fact could have been deduced directly from an inspection of Fig. 6.8, without recourse to equation (6.5), because it is clear that the addition to cycle N of a cycle of lower efficiency (cycle R) could not lead to an increase in efficiency.

The reader should now sketch a temperature-entropy diagram for a $(CICBT)_r$ cycle, (namely, one with intercooling), when it will then be evident that exactly the same considerations apply to this. The $(CICBT)_r$ cycle will be equivalent to a hypothetical simple $(CBT)_r$ cycle N' to which has been added another hypothetical simple $(CBT)_r$ cycle I at the bottom end of cycle N. For this, $\eta_I < \eta_{N'}$, since the pressure ratio is less in cycle I. Hence, as for reheating, the addition of intercooling to a simple $(CBT)_r$ cycle does <u>not</u> improve the thermal efficiency.

6.5 For the special case of a $(CBTRTX)_i$ cycle in which $T_r = T_b$ and $\varepsilon = 1$, show, from first principles, that for maximum efficiency, $p_R = \sqrt{p_A\, p_B}$. The symbols have the same significance as in Section 6.12.

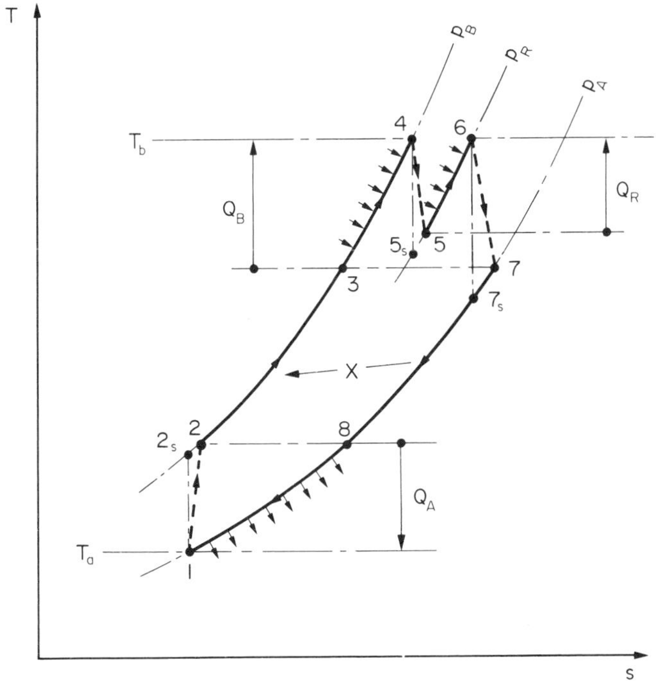

Fig. P.6.5.

Solution

Using the notation in Fig. P.6.5:

Let $\rho \equiv \left(\frac{p_B}{p_A}\right)^{\frac{\gamma-1}{\gamma}}$, $\rho_B = \left(\frac{p_B}{p_R}\right)^{\frac{\gamma-1}{\gamma}}$ and $\rho_R = \left(\frac{p_R}{p_A}\right)^{\frac{\gamma-1}{\gamma}}$

$$\eta_{CY} = 1 - \frac{Q_{out}}{Q_{in}}$$

$$Q_{out} = Q_A = c_p\ (T_8 - T_1)$$

With T_a, T_b, p_A and p_B all constant while p_R is varied, Q_{out} is constant.

∴ η_{CY} is a maximum when Q_{in} is a maximum

$$\frac{Q_{in}}{c_p} = \frac{Q_B + Q_R}{c_p} = (T_4 - T_3) + (T_6 - T_5)$$

But $T_6 = T_4 = T_b$ and $T_3 = T_7$

$$\therefore \frac{Q_{in}}{c_p} = 2T_b - (T_5 + T_7)$$

Hence Q_{in} is a maximum when $(T_5 + T_7)$ is a minimum

For the high - pressure turbine, $(T_b - T_5) = \eta_T\ T_b(1 - \frac{1}{\rho_B})$

For the low - pressure turbine, $(T_b - T_7) = \eta_T\ T_b(1 - \frac{1}{\rho_R})$. But $\rho_R = \frac{\rho}{\rho_B}$

$$\therefore (T_5 + T_7) = T_b\left[2 - \eta_T(1 - \frac{1}{\rho_B}) - \eta_T(1 - \frac{\rho_B}{\rho})\right]$$

$$\therefore \frac{\partial(T_5 + T_7)}{\partial \rho_B} = \eta_T\ T_b\left(-\frac{1}{\rho_B^2} + \frac{1}{\rho}\right)$$

$$= 0 \text{ for a minimum}$$

$$\therefore \rho = \rho_B^2$$

$$\frac{p_B}{p_A} = \frac{p_B^2}{p_R^2}$$

∴ *For maximum* η_{CY}, $p_R = \sqrt{p_A\ p_B}$ $\left(\text{i.e. } \frac{p_B}{p_R} = \frac{p_R}{p_A}\right)$

Note: It has been assumed that η_T is constant and the same for the two turbines.

6.6 Compare the $(CBT)_i$, $(CBTX)_i$ and $(CICBTRTX)_i$ cycles for the conditions given below, by estimating the values of the maximum efficiency, the pressure ratio for maximum efficiency, and the corresponding net work output per kg of fluid circulated. The fluid may be treated as a perfect gas with c_p = 1.01 kJ/kg K and γ = 1.4.

Compressor inlet temperature	15 °C
Turbine inlet temperature	800 °C
Compressor isentropic efficiency	0.85
Turbine isentropic efficiency	0.88
Heat exchanger effectiveness	0.8
Intercooling to	40 °C
Reheating to	800 °C

Solution

As indicated, these optimisation calculations are best made by computer. Readers must therefore be left to write their own computer programmes. Whilst these could allow the computer to perform the necessary optimisation for maximum efficiency in each of the cycles, readers may find it more educationally profitable to arrange for the computer to print out values of W_T, W_C, W_{net} and η_{CY} over a wide range of values of the isentropic temperature ratio, ρ_p. These values can then be plotted and the resulting graphs be compared respectively with those presented in Figures 3.3, 3.5 and 6.5. Note may be taken of the following points:

$(CBT)_i$ cycle

As an alternative to allowing the computer to carry out the necessary optimisation, the optimum value of ρ_p may be determined by differentiating with respect to ρ_p the expression for η_{CY} given and proved in Problem 3.3. In this way, the reader may confirm that the resulting expression for the optimum isentropic temperature ratio of compression, $\rho_{opt.}$, is given by:

$$(\alpha - \beta + 1)\rho^2_{opt.} - 2\alpha\rho_{opt.} + \alpha\beta = 0.$$

The value of $\rho_{opt.}$ may then be calculated, after inserting in this equation the calculated numerical values of α and β for the data given.

$(CICBTRTX)_i$ cycle

When the calculations have been completed, it should be checked that the calculated values of ρ_p, ρ_I and ρ_R satisfy equations (6.9) and (6.10). Here, using the notation of Fig. 6.7:

$$\rho_p \equiv (p_B/p_A)^{(\gamma-1)/\gamma},\ \rho_I \equiv (p_I/p_A)^{(\gamma-1)/\gamma} \text{and } \rho_R \equiv (p_R/p_A)^{(\gamma-1)/\gamma}$$

6.7 In a conceptual Compressed Air Energy Storage scheme, the air pressure in the cavern is maintained constant at p_c during charging and discharging by means of a head of water applied from a water reservoir at a higher level. As a result of heat exchange with its environment, the temperature of the air within the cavern may also be assumed to remain constant at the ambient temperature T_a. To assist in achieving this, there is an aftercooler between the compressor and the cavern. As a means of reducing the required work input during charging, the compression process incorporates intercooling.

During the charging process, an intercooled compressor of isothermal efficiency $\eta_{isoth.}$ is driven by a motor-generator to draw a mass M of air from the atmosphere at p_a, T_a, compress it to p_c and deliver it to the cavern. During the discharging process, the same mass M of air from the cavern is passed through a heat exchanger, where it receives heat from an external source to raise its temperature to T_T at entry to the turbine. The latter exhausts at atmospheric pressure p_a and drives the motor-generator. The isentropic efficiency of the turbine is η_T and the efficiency of the motor-generator is η_{MG} in both the motoring and generating modes. The air may be treated as a perfect gas throughout the process.

Writing $r_p \equiv p_c/p_a$ and $\rho_p \equiv r_p^{(\gamma-1)/\gamma}$, show that the ratio of the electrical output W_G during the discharging process to the electrical input W_M during the charging process is given by the expression

$$\frac{W_G}{W_M} = \eta_{isoth.}\ \eta_T\ \eta_{MG}^2 \left(\frac{\gamma}{\gamma - 1}\right)\left(\frac{\rho_p - 1}{\rho_p}\right)\frac{T_T/T_a}{\ln r_p}.$$

Given that $\eta_{isoth.} = 0.75$, $\eta_T = 0.88$, $\eta_{MG} = 0.97$, $r_p = 45$, $t_T =$ 700 °C and $t_a = 10$ °C, calculate the following quantities:

(a) The ratio W_G/W_M.
(b) The ratio of the net electrical output to the heat input from the external source, expressed as a percentage.
(c) The value of W_G/W_M, and the compressor exit temperature, were the compression process to be adiabatic, with an isentropic efficiency of 88 %.
(d) The temperature at turbine exhaust were no heat to be added to the air between the cavern and the turbine.

Solution

Using the notation of Fig. P.6.7:

Charging process

$$(dW_{in})_{rev} = \int V\,dp = MRT \int \frac{dp}{p}$$

$$\therefore\ (W_{in})_{rev} = MRT_a \ln r_p, \text{ where } r_p \equiv p_c/p_a$$

$$\therefore W_M = \frac{MRT_a}{\eta_{isoth.} \eta_{MG}} \ln r_p \qquad (1)$$

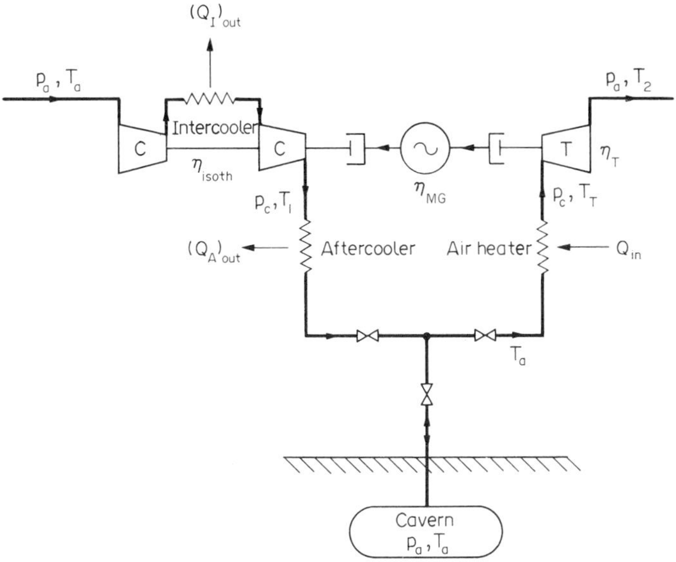

Fig. P.6.7.

Discharging process

$$\frac{T_{2_s}}{T_T} = \frac{1}{\rho_p}, \text{ where } \rho_p \equiv r_p^{(\gamma-1)/\gamma}$$

$$(T_T - T_2) = \eta_T (T_T - T_{2_s}) = \eta_T T_T (1 - \frac{1}{\rho_p})$$

$$\therefore W_G = \eta_{MG} \eta_T M c_p T_T \left(\frac{\rho_p - 1}{\rho_p}\right) \qquad (2)$$

But $R = c_p - c_v$

$$\therefore \frac{c_p}{R} = \frac{c_p/c_v}{(c_p/c_v) - 1} = \frac{\gamma}{\gamma - 1} \qquad (3)$$

Hence, from (1), (2) and (3):

$$\frac{W_G}{W_M} = \eta_{isoth.} \eta_T \eta_{MG}^2 \left(\frac{\gamma}{\gamma - 1}\right)\left(\frac{\rho_p - 1}{\rho_p}\right) \frac{T_T/T_a}{\ln r_p}$$

(a) $r_p = 45$ (given), $\gamma/(\gamma-1) = 1.4/0.4 = 3.5$, $\rho_p = 45^{1/3.5} = 2.9672$

$$\frac{W_G}{W_M} = 0.75 \times 0.88 \times (0.97)^2 \times 3.5 \times \frac{1.9672}{2.9672} \times \frac{973.15}{283.15} \times \frac{1}{3.8067} = \underline{1.30}$$

(b) $$W_G = 0.97 \times 0.88 \times 1.01 \times 973.15 \times \frac{1.9672}{2.9672} M = 556.23 \text{ M kJ}$$

$$W_M = \frac{0.287 \times 283.15}{0.75 \times 0.97} \times 3.8067 \text{ M} = 425.22 \text{ M kJ}$$

$$\therefore W_{net} = (W_G - W_M) = 131.01 \text{ M kJ}$$

$$Q_{in} = M c_p (T_T - T_a) = 1.01 \times 690 \text{ M} = 696.90 \text{ M kJ}$$

$$\therefore \frac{W_{net}}{Q_{in}} = \frac{131.01}{696.90} \times 100 = \underline{18.8}\ \%$$

(c) For adiabatic compression, with $\eta_C = 0.88$:

$$W_M = \frac{M c_p T_a (\rho_p - 1)}{\eta_C \eta_{MG}} = \frac{1.01 \times 283.15 \times 1.9672}{0.88 \times 0.97} M = 659.07 \text{ M kJ}$$

$W_G = 556.23$ M kJ (as before)

$$\therefore \frac{W_G}{W_M} = \frac{556.23}{659.07} = \underline{0.844}$$

$$T_1 - T_a = \frac{T_a(\rho_p - 1)}{\eta_C} = \frac{283.15 \times 1.9672}{0.88} = 633.0 \text{ K}$$

$\therefore$ Compressor exit temperature, $t_1 = t_a + (T_1 - T_a)$

$$= 10 + 633.0 = \underline{643}\ °\text{C}$$

(d) If $T'_T = 283.15$ K and $\rho_p = 2.9672$

$$(T'_T - T'_2) = T'_T \left(\frac{\rho_p - 1}{\rho_p}\right) = 283.15 \times \frac{1.9672}{2.9672} = 187.7 \text{ K}$$

$\therefore$ Turbine exhaust temperature, $t'_2 = t'_T - (T'_T - T'_2)$

$$= (10 - 187.7) \approx \underline{-178}\ °\text{C}$$

6.8 The compression process in Problem 6.7 is carried out in four adiabatic stages in series, with the same pressure ratio across each and with intercooling between stages. Assuming that the air is cooled to ambient temperature T_a at exit from each intercooler, and that the isentropic efficiency η_C is the same for all four stages, determine the value of η_C corresponding to the given value of η_{isoth}. Neglect parasitic pressure drops.

Note: The isothermal efficiency of a compressor is defined as the ratio of the ideal work input for reversible, isothermal compression over the given pressure ratio to the actual work input to the compressor.

Solution

$$\text{Pressure ratio per stage} \equiv r_p' = 45^{1/4} = 2.5900$$

$$\rho_p' \text{ per stage} = (2.5900)^{1/3.5} = 1.3125$$

$$\text{Actual total work input} = 4 \times \frac{M\ c_p\ T_a\ (\rho_p' - 1)}{\eta_C}$$

$$\text{Reversible isothermal work input} = M\ R\ T_a\ \ln r_p, \text{ where } r_p = 45$$

$$\eta_{isoth.} \equiv \frac{\text{Rev. isothermal work input}}{\text{Actual total work input}} = \frac{\eta_C\ M\ R\ T_a\ \ln r_p}{4\ M\ c_p\ T_a\ (\rho_p' - 1)}$$

$$\text{But } c_p/R = \gamma/(\gamma-1)$$

$$\therefore \eta_C = \frac{4\gamma}{\gamma-1}\ \frac{(\rho_p' - 1)}{\ln r_p}\ \eta_{isoth.}$$

$$= \frac{4 \times 3.5 \times 0.3125}{3.8067} \times 0.75 = \underline{0.862}$$

Problems 8.7 and 8.8 in Chapter 8, relating to gas-turbine cycles for nuclear power plant, provide further practice in gas-turbine cycle calculations.

CHAPTER 7

Advanced Steam-Turbine Plant

7.1 Show that the flow rate through the boiler per unit flow rate through the condenser for the hypothetical ideal plant described in Section 7.5 is equal to $\Delta s_1/\Delta s_2$, where these quantities are defined as in Fig. 7.4.

Calculate this ratio when the boiler steam conditions are 10 MN/m², 550 °C and the condenser pressure is 7 kN/m².

Solution

Using the notation of Fig. 7.4(b):

From equation (7.3), $$\eta_{CY} = \frac{b_3 - b_2}{h_3 - h_2} = \frac{(h_3 - h_2) - T_A(s_3 - s_2)}{(h_3 - h_2)}$$

$$\therefore\ (1 - \eta_{CY}) = \frac{T_A(s_3 - s_2)}{h_3 - h_2} \qquad (1)$$

$$\text{But}\ (1 - \eta_{CY}) = \frac{Q_{out}}{Q_{in}} = \frac{M_A(h_5 - h_1)}{M_B(h_3 - h_2)} = \frac{M_A T_A(s_5 - s_1)}{M_B(h_3 - h_2)} \qquad (2)$$

Equating (1) and (2), $$\frac{M_B}{M_A} = \frac{s_5 - s_1}{s_3 - s_2} = \frac{\Delta s_1}{\Delta s_2}$$

At state 3 (10 MN/m², 550 °C) : $s_3 = 6.756$ kJ/kg K

At state 2 (Sat. liquid at 10 MN/m²): $s_2 = 3.361$ kJ/kg K

At state 1 (Sat. liquid at 7 kN/m²): $s_1 = 0.559$ kJ/kg K

$$\therefore\ \frac{M_B}{M_A} = \frac{6.756 - 0.559}{6.756 - 3.361} = \frac{6.197}{3.395} = \underline{1.825}$$

7.2 For the hypothetical ideal plant discussed in Section 7.6, sketch a temperature-entropy diagram, similar to that in Fig. 7.5(b), for an infinitesimal feed-heating stage supplied from a point in the turbine at which the steam is superheated.

Solution

As noted in Section 7.6, in order to ensure full reversibility, it would be necessary for a drain-water turbine to be included in each line taking the condensed bled steam down from each surface heater to the preceding heater in the feed line, as depicted in Fig. 7.6. Also, if the steam bled to a feed heater were superheated, a multi-stage, intercooled compressor would be needed in the bled steam line from the turbine to the heater, as depicted in Fig. 7.4 and Fig. 7.5.

Thus the flow diagram for such an infinitesimal stage, and the corresponding temperature-entropy diagram, would be as depicted in Fig. P.7.2.

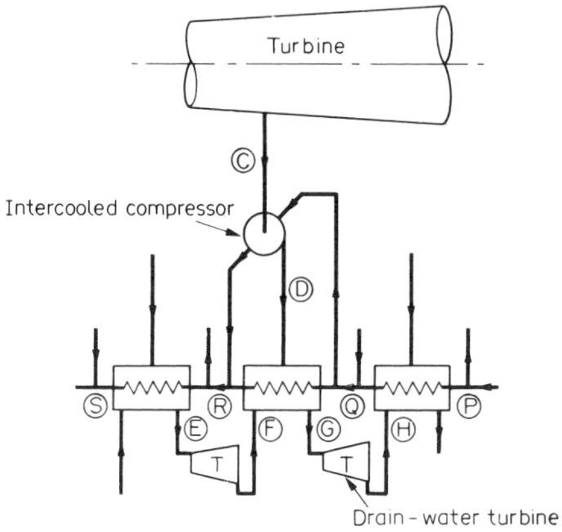

(a) Flow diagram

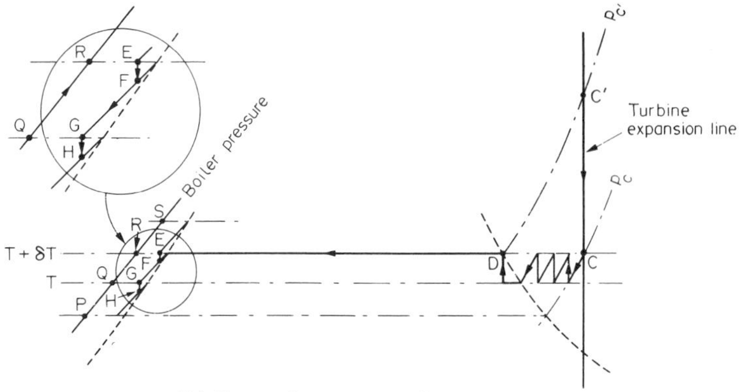

(b) Temperature-entropy diagram

Fig. P.7.2.

7.3 Steam is extracted from a turbine at three points to supply bled steam to a train of feed heaters. The steam conditions at the respective points in the turbine at the design load are as follows:

Position	Inlet	No. 3 heater	No. 2 heater	No. 1 heater	Exhaust
Pressure (kN/m^2)	4500	430	110	34	3.0
Temperature (°C)	440	190	-	-	-
Dryness fraction	-	-	0.978	0.937	0.876

Plot the turbine expansion line on an enthalpy-entropy diagram, and on the same diagram draw a line of constant β = 2180 kJ/kg (see Section 7.11).

Solution

$\beta \equiv (H - h)$
where $H \equiv$ specific enthalpy of steam at a given pressure
and $h \equiv$ specific enthalpy of saturated water at the same pressure

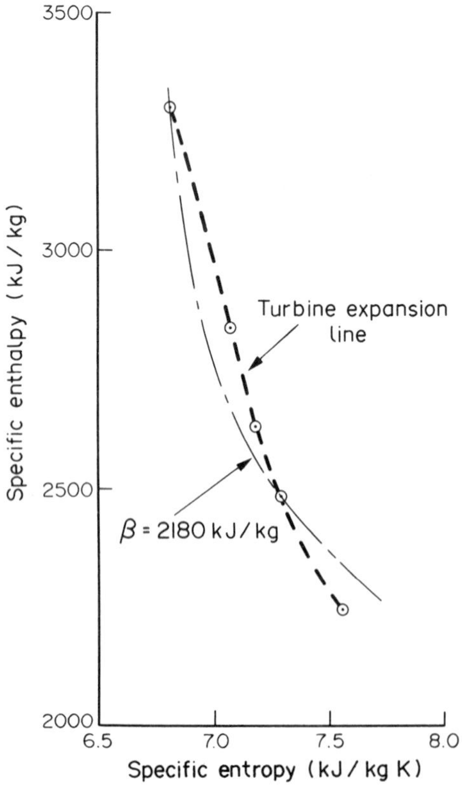

Fig. P.7.3.

7.4 Calculate the exact Rankine cycle efficiency (taking due account of the feed pump work) and the corresponding heat rate in Btu/kW h when the steam is supplied to the turbine at 6 MN/m² and 500 °C, and exhausts to the condenser at 4 kN/m².

In an ideal, reversible regenerative cycle operating with the same steam conditions, the feed water is raised to the boiler saturation temperature in an infinite number of feed-heating stages. Calculate the cycle efficiency and heat rate, and the percentage reduction in heat rate due to feed-heating.

Solution

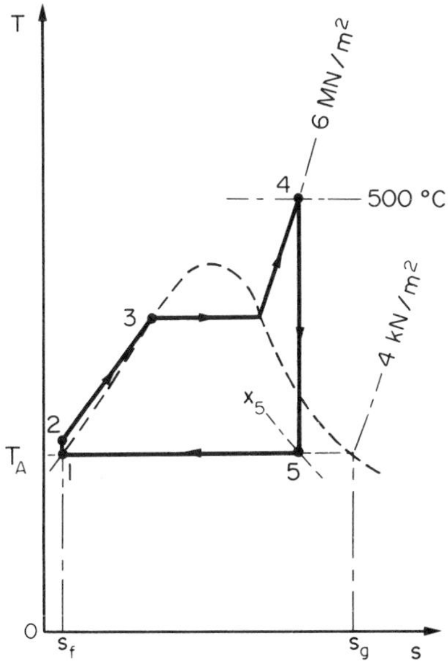

Fig. P.7.4.

Rankine cycle

At 6 MN/m² and 500 °C:

h_4 = 3422 kJ/kg, s_4 = 6.882 kJ/kg K

At 4 kN/m²:

s_g = 8.475 kJ/kg K, s_f = 0.422 kJ/kg K

h_g = 2554.5 kJ/kg, h_{fg} = 2433.1 kJ/kg

Dryness fraction, x_5, at state 5:

$$(1 - x_5) = \frac{8.475 - 6.882}{8.475 - 0.422} = \frac{1.593}{8.053} = 0.1978$$

$$h_5 = 2554.5 - 0.1978 \times 2433.1 = 2073.2 \text{ kJ/kg}$$

Turbine:

$$W_T = 3422 - 2073 = 1349 \text{ kJ/kg}$$

Feed pump:

$$W_P = v_f(p_2 - p_1) = \frac{0.001004 \times 5996 \times 10^3}{10^3} = 6.0 \text{ kJ/kg}$$

$$\therefore W_{net} = W_T - W_P = 1343 \text{ kJ/kg}$$

$$h_2 = h_f + W_P = 121.4 + 6.0 = 127.4 \text{ kJ/kg}$$

$$Q_{in} = h_4 - h_2 = 3422 - 127 = 3295 \text{ kJ/kg}$$

$$\therefore \text{Exact } \eta_{RANK} \equiv \frac{W_{net}}{Q_{in}} = \frac{1343}{3295} \times 100 = \underline{40.76} \ \%$$

$$\text{Heat rate} \equiv \frac{Q_{in}}{W_{net}} = \frac{1}{\eta_{CY}} \ [(\text{kW h})/(\text{kW h})]$$

But 1 kW h ≈ 3412 Btu

$$\therefore \text{Heat rate} = \frac{3412}{0.4076} = \underline{8371}$$ (Btu of heat input per kW h of work output)

<u>Ideal, reversible regenerative cycle</u>

With the feed water raised to boiler saturation temperature (state 3):

$$h_3 = 1213.7 \text{ kJ/kg}, \quad s_3 = 3.027 \text{ kJ/kg K}$$

$$t_A = 29.0 \text{ °C}, \quad T_A = 302.15 \text{ K}$$

$$\text{Ideal } \eta_{CY} = \frac{b_4 - b_3}{h_4 - h_3} = \frac{(h_4 - h_3) - T_A(s_4 - s_3)}{(h_4 - h_3)}$$

$$= \frac{(3422 - 1213.7) - 302.15(6.882 - 3.027)}{(3422 - 1213.7)}$$

$$= \frac{2208.3 - 1164.8}{2208.3}$$

$$\therefore \text{Ideal } \eta_{CY} = \frac{1043.5}{2208.3} \times 100 = \underline{47.25} \ \%$$

$$\text{Heat rate} = \frac{3412}{0.4725} = \underline{7221} \text{ Btu/kW h}$$

$$\text{Percentage reduction in heat rate} = \frac{8371 - 7221}{8371} \times 100 = \underline{13.74} \ \%$$

7.5 In a hypothetical cyclic steam power plant incorporating a single direct-contact feed heater, the steam leaves the boiler at 1 MN/m² and 400 °C and the pressure in the condenser is 3.5 kN/m². Expansion in the turbine is reversible and adiabatic. The feed heater takes steam bled from the turbine at a pressure of 70 kN/m² and heats the feed water to the corresponding saturation temperature. Temperature and enthalpy changes of the fluid in passing through any pumps may be neglected. Calculate:

(1) the mass of steam bled from the turbine per kilogram of steam leaving the boiler;
(2) the thermal efficiency of the cycle;
(3) the improvement in thermal efficiency due to the introduction of this single stage of feed heating, expressed as a percentage of the Rankine cycle efficiency.

Solution

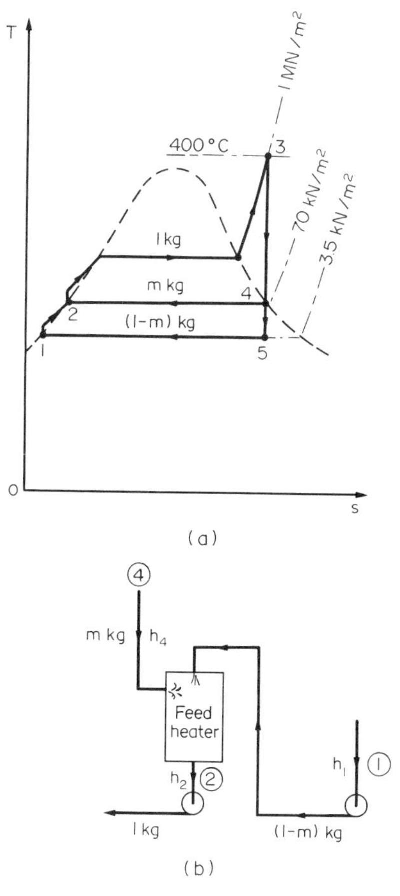

Fig. P.7.5.

Using the enthalpy-entropy diagram and Tables of Ref. 2:

h_1 = 111.8 kJ/kg s_1 = 0.391 kJ/kg K

h_2 = 376.8 s_2 = 1.192

h_3 = 3264 s_3 = 7.467

h_4 = 2657 s_4 = 7.467

h_5 = 2234 s_5 = 7.467

(1) Steady-flow energy equation for the feed heater:

$$h_2 = m\,h_4 + (1 - m)\,h_1$$

$$\therefore\ m = \frac{h_2 - h_1}{h_4 - h_1} = \frac{265.0}{2545.2} = \underline{0.1041}\ \text{kg}$$

(2) Thermal efficiency of the cycle:

$$Q_{in} = (h_3 - h_2) = 2887\ \text{kJ}$$

$$Q_{out} = (1 - m)(h_5 - h_1)$$

$$= 0.8959 \times 2122.2 = \underline{1901}\ \text{kJ}$$

$$\therefore\ W_{net} = 986\ \text{kJ}$$

$$\therefore\ \text{Cycle efficiency, } \eta_{CY} = \frac{986}{2887} \times 100 = \underline{34.2}\ \%$$

(3) Rankine cycle efficiency:

$$Q_{in} \approx (h_3 - h_1) = 3152\ \text{kJ}$$

$$Q_{out} = (h_5 - h_1) = \underline{2122}\ \text{kJ}$$

$$W_{net} = 1030\ \text{kJ}$$

$$\therefore\ \text{Rankine cycle efficiency, } \eta_{RANK} = \frac{1030}{3152} \times 100 = \underline{32.7}\ \%$$

Percentage improvement due to feed heating:

$$\text{Improvement} = \frac{34.2 - 32.7}{32.7} \times 100 = \underline{4.6}\ \%$$

7.6 For the same conditions at boiler inlet and exit as in Problem 7.5, and the same condenser pressure, calculate the thermal efficiency of an internally reversible, regenerative steam cycle in which heat is rejected reversibly to the environment, which is at a temperature T_0 equal to the steam saturation temperature in the condenser. Thence determine the resulting percentage improvement in thermal efficiency, expressed as a percentage of the Rankine cycle efficiency.

Show that the thermal efficiency of this internally reversible , regenerative steam cycle is equal to $[1 - (T_0/\overline{T}_B)]$, where $\overline{T}_B$ is the mean temperature of heat reception on the temperature-entropy diagram.

Calculate $\overline{T}_B$ for this cycle and for the Rankine cycle. Thence, verify that the improvement in thermal efficiency resulting from the introduction of the reversible feed heating process is due to raising the mean temperature of heat reception.

Solution

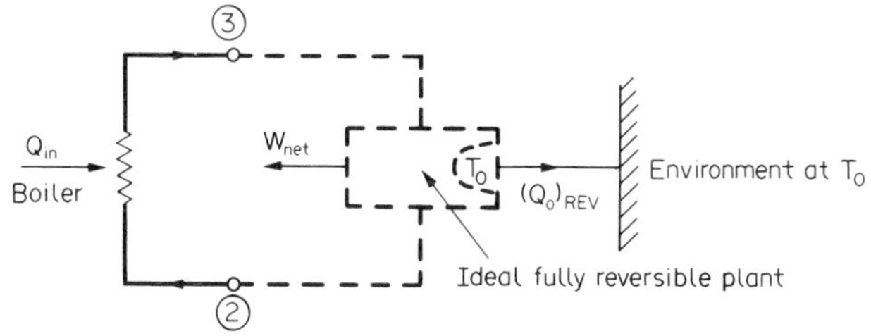

Fig. P.7.6(a).

$$T_0 = 26.7 + 273.15 = 299.85 \text{ K}$$

Using property values from Problem 7.5:

$$\text{Ideal } W_{net} = [W_{REV}]_3^2 = (b_3 - b_2), \text{ where } b \equiv h - T_0 s$$
$$= (h_3 - h_2) - T_0(s_3 - s_2)$$
$$= 2887 - 299.85 \times 6.275$$
$$= 1005 \text{ kJ/kg}$$
$$Q_{in} \approx (h_3 - h_2) = 2887 \text{ kJ/kg}$$
$$\therefore \text{ Ideal } \eta_{CY} = \frac{1005}{2887} \times 100 = \underline{34.8} \%$$

Percentage improvement in thermal efficiency (relative to Rankine cycle)

$$\text{Improvement} = \frac{34.8 - 32.7}{32.7} \times 100 = \underline{6.4} \%$$

Ideal regenerative cycle [Fig. P.7.6(b).]

$$W_{net} = (b_3 - b_{2_s}) = (b_3 - b_2), \text{ neglecting enthalpy rise in pump}$$
$$\therefore W_{net} = (h_3 - h_2) - T_0(s_3 - s_2)$$
$$= Q_{in} - Q_{out}$$
$$\text{Ideal } \eta_{CY} = \left(1 - \frac{Q_{out}}{Q_{in}}\right),$$

where $Q_{out} = T_0(s_3 - s_2)$ (1)

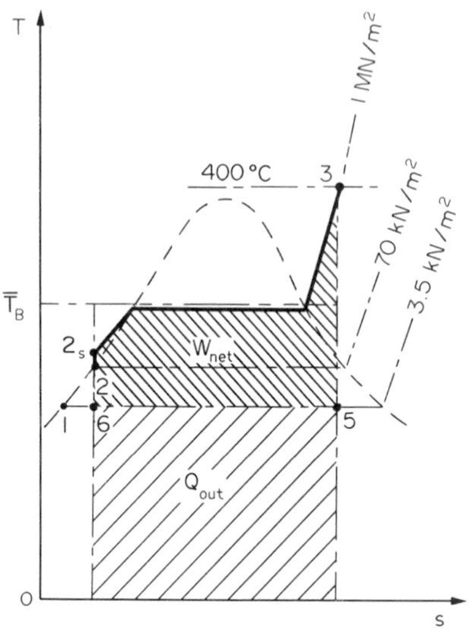

Fig. P.7.6(b).

and $Q_{in} \approx (h_3 - h_2) = \overline{T}_B(s_3 - s_2)$ (2)

since, by definition, $\overline{T}_B \equiv \dfrac{\int_2^3 T\,ds}{s_3 - s_2} = \dfrac{h_3 - h_2}{s_3 - s_2}$ (3)

Whence, from (1) and (2),

$$\textit{Ideal}\ \eta_{CY} = \left(1 - \frac{Q_{out}}{Q_{in}}\right) = \left(1 - \frac{T_0}{\overline{T}_B}\right)$$

[Note that W_{net} and Q_{out} are thus represented by the shaded areas in Fig. P.7.6(b)].

Thus, mean temp. of heat reception, $\overline{T}_B = \dfrac{h_3 - h_2}{s_3 - s_2} = \dfrac{2887}{6.275} = \underline{460.1}$ K

Ideal $\eta_{CY} = \left(1 - \dfrac{T_0}{\overline{T}_B}\right) = \left(1 - \dfrac{299.85}{460.1}\right) \times 100 = \underline{34.8}$ %*

Rankine cycle

Mean temp. of heat reception, $\overline{T}_B = \dfrac{h_3 - h_1}{s_3 - s_1} = \dfrac{3152}{7.076} = \underline{445.4}$ K

$\eta_{RANK} = \left(1 - \dfrac{T_0}{\overline{T}_B}\right) = \left(1 - \dfrac{299.85}{445.4}\right) \times 100 = \underline{32.7}$ %*

*The fact that these are the same as the values previously calculated

confirms that the improvement in thermal efficiency resulting from the introduction of the reversible feed heating process is due to the raising of the mean temperature of heat reception.

7.7 In Problem 7.5, calculate the entropy creation due to irreversibility in the feed heater (namely, the net entropy increase of the fluid streams in passing through the feed heater). Thence evaluate the loss of gross work output due to the irreversibility in the feed heater, taking the environment temperature T_0 as being equal to the steam saturation temperature in the condenser (see Theorem 2 in Section A.4 of Appendix A). Express this lost work as a percentage of the net work output in the cycle and verify that this percentage is approximately equal to the difference between the percentage improvements in thermal efficiency in Problems 7.5 and 7.6.

Solution

By definition, entropy creation, ΔS_C, due to irreversibility is given by

$$\Delta S_C \equiv \Delta S - \Delta S_Q.$$

In the present context, ΔS is the net entropy increase of the fluid streams in their passage through the specified control volume (in this case, the feed heater), while ΔS_Q is the net thermal entropy flux entering the control volume due to heat transfers across its boundary. (See Section A.6 in Appendix A of Ref. 2 and Sections 12.7 to 12.12 in Chapter 12 of Ref. 3)

As stray heat loss from the feed heater is assumed to be negligible,

$$\Delta S_Q = 0.$$

Hence, using the notation of Fig. P.7.5(b):

$$\Delta S_C = \Delta S = s_2 - [m\, s_4 + (1 - m)\, s_1]$$

$$= (s_2 - s_1) - m\,(s_4 - s_1)$$

$\therefore$ Entropy creation due to irreversibility, $\Delta S_C = 0.801 - 0.1041 \times 7.076$

$= \underline{0.064}$ kJ/K

Loss of gross work output due to irreversibility,

$$[(W_g)_R - (W_g)_I] = T_0\, \Delta S_C \quad \text{[Equation (A.14) in Appendix A of Ref. 1]}$$

$$= 299.85 \times 0.064 = \underline{19.2} \text{ kJ}$$

Lost work as a percentage of $W_{net} = \dfrac{19.2}{986} \times 100 = \underline{1.9}$ %

Allowing for rounding errors, this value agrees with the difference of 1.8 % between the values of 6.4 % and 4.6 % calculated respectively in Problems 7.6 and 7.5.

Attention is drawn to the fact that, of the theoretical gain of about 6½ % ideally obtainable as a result of raising the mean temperature of heat reception in the cycle through the introduction of feed heating, there is a loss of about 2 % due to the irreversibility of the mixing process in the feed heater, giving a net gain of thermal efficiency of about 4½ % due to feed heating.

Note: In Problems 7.8 and 7.9, the work input to all pumps may be neglected, the enthalpy of water at any temperature being taken as equal to the saturation enthalpy at that temperature.

7.8 In a regenerative steam plant the turbine inlet conditions and exhaust pressure are as in Problem 7.4, but the feed water is raised to a temperature of 182 °C in four heaters.

(a) If direct-contact heaters were used throughout, each raising the feed water to the saturation temperature of the steam supplied to the heater, and if the overall enthalpy rise of the feed water were divided equally amongst the heaters, what would be the required bled steam pressures?

(b) In these circumstances, the bled steam enthalpies would be respectively 3040, 2868, 2683 and 2490 kJ/kg, and the enthalpy of the steam at turbine exhaust 2285 kJ/kg. Calculate the steam flow rate through the boiler per unit flow rate to the condenser. What would be the calculated value of this ratio if the values of β for all heaters were taken to be equal to the mean value? Determine the cycle efficiency and heat rate.

(c) Calculate the cycle efficiency and heat rate (in Btu/kW h) for a non-regenerative cycle in which the states of the steam at turbine inlet and exhaust are the same as in this regenerative cycle. Thence determine the percentage reduction in heat rate due to feed heating. From this result and that calculated in Problem 7.4 determine the values of x and y, as defined respectively in Sections 7.12 and 7.13, and plot this point on Fig. 7.9.

Solution

At No. 4 heater exit, temperature = 182 °C, $h = h_f$ = 772.0 kJ/kg

At condenser exit, pressure = 4 kN/m², $h = h_f$ = 121.4 kJ/kg

Overall enthalpy rise = 650.6 kJ/kg

∴ Enthalpy rise per heater = 650.6/4 = 162.65 kJ/kg

The bled steam pressures to the heaters will be the saturation pressures corresponding to the values of h_f at exit from the heaters. Hence:

	Heater No.		1	2	3	4
	h_f	kJ/kg	284.1	446.7	609.4	772.0
(a)	Pressure	MN/m²	*0.0284*	*0.127*	*0.412*	*1.050*
	H	kJ/kg	2490	2683	2868	3040
	$h = h_f$	kJ/kg	284	447	609	772
	$\beta \equiv (H - h)$	kJ/kg	2206	2236	2259	2268

$$\beta_{mean} = 2242 \quad kJ/kg$$

Enthalpy rise per heater, r = 162.65 kJ/kg

r/β	0.0737	0.0727	0.0720	0.0717
$\gamma \equiv \left(1 + \frac{r}{\beta}\right)$	1.0737	1.0727	1.0720	1.0717

(b) $\dfrac{\text{Steam flow from boiler}}{\text{Steam flow to condenser}} = \prod_{1}^{4} \gamma = \underline{1.323}$

Alternatively, $\gamma_{mean} = \left(1 + \dfrac{r}{\beta_{mean}}\right) = \left(1 + \dfrac{162.65}{2242}\right) = 1.0725$

$\dfrac{\text{Steam flow from boiler}}{\text{Steam flow to condenser}} = \gamma_{mean}^{4} = (1.0725)^{4} = \underline{1.323}$

At boiler exit (6 MN/m^2, 500 °C), specific enthalpy = 3422 kJ/kg

At condenser exit (4 kN/m^2), specific enthalpy = 121.4 kJ/kg

$Q_{in} = 1.323\ (3422 - 772) = 3506.0$ kJ

$Q_{out} = (2285 - 121.4) = \underline{2163.6}$ kJ

$W_{net} = 1342.4$ kJ

∴ Cycle efficiency, $\eta_{CY} = \dfrac{1342.4}{3506.0} \times 100 = \underline{38.3}\ \%$

But 1 kW h ≈ 3412 Btu

∴ Heat rate $= \dfrac{3412}{0.383} = \underline{8910}$ Btu/kW h

(c) Non-regenerative cycle with the same turbine isentropic efficiency

Neglecting work input to feed pump:

$Q_{in} = 3422 - 121.4 = 3300.6$ kJ/kg

$Q_{out} = \underline{2163.6}$ kJ/kg (as before)

$W_{net} = 1137.0$ kJ/kg

[Check: $W_{net} \approx W_T = 3422 - 2285 = 1137$ kJ/kg]

∴ Cycle efficiency, $\eta_{CY} = \dfrac{1137.0}{3300.6} \times 100 = \underline{34.45}\ \%$

Heat rate $= \dfrac{3412}{0.3445} = \underline{9904}$ Btu/kW h

Reduction in heat rate = 9904 - 8910 = 994 Btu/kW h

Percentage reduction in heat rate $= \dfrac{994}{9904} \times 100 = \underline{10.04}\ \%$

From the definitions of x and y in Sections 7.12 and 7.13 of Ref. 1:

$$x \equiv \frac{\text{Total enthalpy rise of feed water}}{\text{Maximum possible enthalpy rise}}$$

h_f at boiler pressure (6 MN/m²) = 1213.7 kJ/kg

$$\therefore\ x = \frac{772 - 121.4}{1213.7 - 121.4} = \frac{650.6}{1092.3} = \underline{0.596}$$

$$y \equiv \frac{C_0 - C}{C_0 - C_\infty},$$

where C_0, C and C_∞ are respectively the heat rates when there is no feed heating (i.e. Rankine cycle), when heating with four heaters to the specified temperature (as in this Problem), and when feed heating with an infinite number of heaters to the boiler saturation temperature (as in Problem 7.4).

From the above calculations, $C_0 - C \propto 10.04\ \%$

From Problem 7.4, $C_0 - C_\infty \propto 13.74\ \%$

$$\therefore\ y = \frac{10.04}{13.74} = \underline{0.731}$$

When the point (x,y) is plotted on Fig. 7.9, it is seen to lie correctly on the curve for 4 heaters.

7.9 In a regenerative steam plant the states of the steam at turbine inlet and exhaust, and at the bled points, are the same as in Problem 7.8, but the feed water from the condenser passes first through two direct-contact heaters and then through two surface heaters. In the surface heaters there is a temperature difference of 5 K between the saturation temperature of the bled steam supplied to a heater and the outlet temperature of the feed water from that heater. The condensed bled steam leaves a surface heater at the saturation temperature of the steam supplied to the heater, and these drains from the surface heaters are cascaded successively from heater to heater, passing finally to the direct-contact heater preceding the surface heaters.

Calculate the flow rate through the boiler per unit flow rate to the condenser, the cycle efficiency and heat rate (in Btu/kW h). (Hint: First determine the condenser flow rate per unit flow rate through the boiler, starting the calculation of the bled steam quantities at the heater nearest the boiler).

Solution

Using the notation in the accompanying Fig. P.7.9:

H_c = 2285 kJ/kg	h_c = 121.4 kJ/kg
H_1 = 2490	h_1 = 284.1
H_2 = 2683	h_2 = 446.7
H_3 = 2868	h_3 = 609.4
H_4 = 3040	h_4 = 772.0

$$h_3' = 587.6 \text{ kJ/kg}$$

$$h_4' = 749.7 \text{ kJ/kg}$$

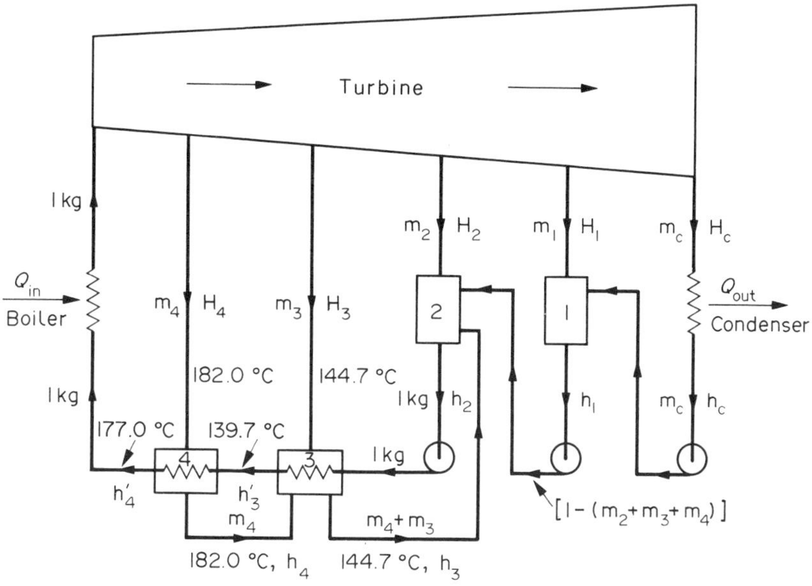

Fig. P.7.9.

Per unit mass flow to the boiler, the respective bled steam flows to the heaters are calculated from the steady-flow energy equations for the heaters, as follows:

Heater 4

$$m_4 = \frac{h_4' - h_3'}{H_4 - h_4} = \frac{749.7 - 587.6}{3040 - 772.0} = 0.0715 \text{ kg}$$

Heater 3

$$m_3(H_3 - h_3) + m_4(h_4 - h_3) = 1 \times (h_3' - h_2)$$

$$2258.6\ m_3 + 0.0715 \times 162.6 = 140.9$$

$$m_3 = \frac{129.3}{2258.6} = 0.0572 \text{ kg}$$

$$m_3 + m_4 = 0.1287$$

$$[1 - (m_3 + m_4)] = 0.8713$$

Heater 2

$$m_2(H_2 - h_2) + (m_3 + m_4)(h_3 - h_2) = [1 - (m_2 + m_3 + m_4)](h_2 - h_1)$$

$$2236.3\ m_2 + 0.1287 \times 162.7 = (0.8713 - m_2) \times 162.6$$

$$m_2(2236.3 + 162.6) = 141.67 - 20.94$$

$$m_2 = \frac{120.73}{2398.9} = 0.0503 \text{ kg}$$

$$(m_2 + m_3 + m_4) = 0.1790$$

$$[1 - (m_2 + m_3 + m_4)] = 0.8210$$

Heater 1

$$m_1(H_1 - h_1) = (0.8210 - m_1)(h_1 - h_c)$$

$$2205.9\ m_1 = (0.8210 - m_1) \times 162.7$$

$$m_1 = \frac{133.58}{2368.6} = 0.0564 \text{ kg}$$

Steam flow to condenser $= m_c = (0.8210 - m_1) = 0.7646$ kg

$$\therefore \frac{\text{Boiler flow}}{\text{Condenser flow}} = \frac{1}{0.7646} = \underline{1.308}$$

Cycle efficiency and heat rate

$$Q_{in} = 1.308\ (3422 - 749.7) = 3495.4 \text{ kJ}$$

$$Q_{out} = (2285 - 121.4) = 2163.6 \text{ kJ}$$

$$W_{net} = 1331.8 \text{ kJ}$$

$$\therefore \eta_{CY} = \frac{1331.8}{3495.4} \times 100 = \underline{38.1}\ \%$$

$$\text{Heat rate} = \frac{3412}{0.381} = \underline{8955} \text{ Btu/kW h}$$

7.10 In Fig. 7.9, it is seen that, for any given number of heaters n, the fractional improvement in heat rate when heating to the boiler saturation temperature (i.e. when x = 1) is the same as the optimum fractional improvement for (n - 1) heaters. Explain why this is so.

Solution

It was shown in Section 7.12 that, for maximum efficiency in a non-reheat plant, the enthalpy rises should, to a first approximation, be the same in all heaters and in the economiser (which is taken to heat the feed water to boiler saturation temperature from its temperature at exit from the last heater). Figure 7.9 is based on that approximation.

Now, if there are n heaters, of which the last heater itself heats the feed water to the boiler saturation temperature (thus resulting in no need for an economiser section in the boiler), this last heater will need to take its bled steam direct from boiler exit, at boiler pressure. So the effect will be exactly the same as if this last enthalpy rise had taken place in the boiler, instead of in the last heater. The heating in this last heater can therefore produce no gain in cycle efficiency due to feed heating.

Given that all enthalpy rises in the heaters are taken to be equal (for maximum efficiency), this last heater will provide 1/n th of the total enthalpy rise from the saturation enthalpy at condenser pressure to the saturation enthalpy at boiler pressure. Thus, the gain in efficiency when n heaters are used to heat the feed water to boiler saturation temperature will be the same as that obtained by heating the feed water in (n - 1) heaters over a fractional enthalpy rise, x, equal to (n - 1)/n. This is the same as the optimum fractional enthalpy rise, x_{opt}, for (n - 1) heaters, as may be seen by replacing n by (n - 1) in equation (7.15), namely $x_n = n/(n + 1)$. To take a specific case, namely 3 heaters heating to boiler saturation temperature (at x = 1), the gain in efficiency will be the same as (n - 1) = 2 heaters heating to x = [(n - 1)/n] = 2/3. As shown in Fig. 7.9, this is equal to the optimum fractional enthalpy rise for 2 heaters. The given proposition is thus proved.

The original and fuller analysis of this matter is given in a paper by the author which appeared in Proc. I. Mech. E., vol. 161, pp. 157-164, in 1949 under the title - "A Generalized Analysis of the Regenerative Steam Cycle for a Finite Number of Heaters".

7.11 In an ideal, reversible non-regenerative steam cycle the initial steam pressure and temperature are respectively 15 MN/m² and 500 °C and the condenser pressure is 4 kN/m².

Calculate the percentage exhaust wetness and the cycle efficiency (1) when there is no reheating, and (2) when the steam is reheated to 500 °C at (a) 6 MN/m², (b) 4 MN/m², (c) 2 MN/m². The work input to the feed pump may not be neglected.

From these results express the optimum reheat pressure as a fraction of the initial pressure, and determine the percentage improvement in efficiency due to reheating at this pressure. Use equation (7.18) to check the calculated improvement.

Note: In obtaining the given answers, use was made of the tabulated data given in the publication - UK Steam Tables in SI Units 1970, Ed. Arnold (Publishers) Ltd., London (1970). Those Tables have also been used here.

Solution

(1) When there is no reheating (Rankine cycle)

1 MN/m² ≡ 10 bar

At 15 MN/m² (150 bar) and 500 °C:

h_3 = 3310.6 kJ/kg, s_3 = 6.3487 kJ/kg K

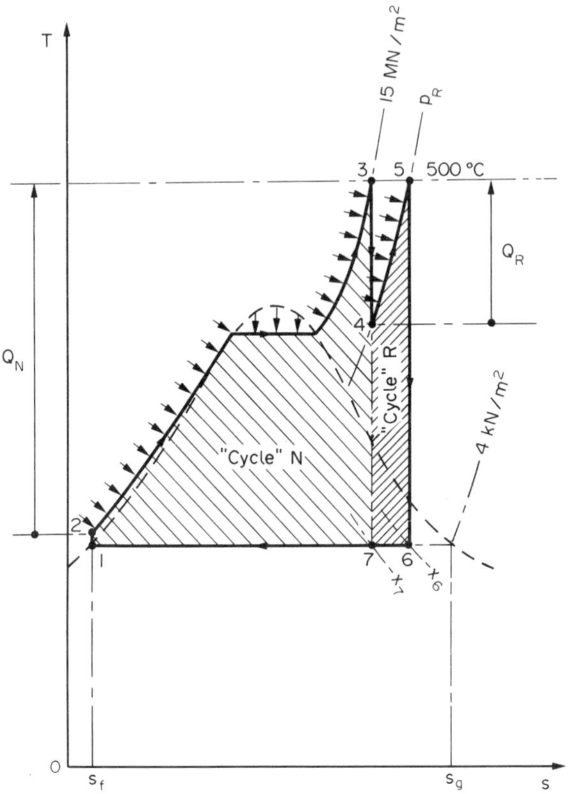

Fig. P.7.11.

At 4 kN/m² (0.04 bar):

s_g = 8.4755 kJ/kg K, s_f = 0.4225 kJ/kg K

h_g = 2554.5 kJ/kg, h_f = 121.4 kJ/kg

Dryness fraction, x_7, at state 7:

$$(1 - x_7) = \frac{8.4755 - 6.3487}{8.4755 - 0.4225} = \frac{2.1268}{8.0530} = 0.2641$$

$\therefore$ Percentage exhaust wetness = *26.4* %

h_7 = 2554.5 - 0.2641 x 2433.1 = 1911.9 kJ/kg

Q_{out} = $(h_7 - h_1)$ = 1911.9 - 121.4 = 1790.5 kJ/kg

At state 2: p_2 = 150 bar, s_2 = 0.4225 kJ/kg K

From Table 3 of the UK Steam Tables, h_2 = 136.5 kJ/kg

$Q_{in} = (h_3 - h_2) = 3310.6 - 136.5 = 3174.1$ kJ/kg

$W_{net} = Q_{in} - Q_{out} = 3174.1 - 1790.5 = 1383.6$ kJ/kg

∴ Cycle efficiency, $\eta_{CY} = \frac{1383.6}{3174.1} \times 100 = \underline{43.59\ \%}$

(2a) With a reheat pressure, p_R, of 6 MN/m²

Using the notation of Fig. P.7.11, and data from Table 3 of the UK Steam Tables:

At 6 MN/m² (60 bar) and 500 °C:

$h_5 = 3422.2$ kJ/kg, $s_5 = 6.8818$ kJ/kg K

At 4 kN/m² (0.04 bar):

$s_g = 8.4755$ kJ/kg K, $s_f = 0.4225$ kJ/kg K

$h_g = 2554.5$ kJ/kg, $h_f = 121.4$ kJ/kg

Dryness fraction, x_6, at state 6:

$$(1 - x_6) = \frac{8.4755 - 6.8818}{8.4755 - 0.4225} = \frac{1.5937}{8.0530} = 0.1979$$

∴ Percentage exhaust wetness = $\underline{19.8\ \%}$

Calculation of cycle efficiency, η_{CY}

$s_4 = s_3 = 6.3487$ kJ/kg K, $p_4 = 6$ MN/m² (60 bar).

Thus, the steam is superheated at state 4, and $t_4 = 352.2$ °C

Hence, $h_4 = 3052.1$ kJ/kg, $h_5 = 3422.2$ kJ/kg

Thus, in the reheater, $Q_{in} = (h_5 - h_4) = 370.1$ kJ/kg

As before, in the boiler proper, $Q_{in} = \underline{3174.1}$ kJ/kg

∴ Total $Q_{in} = 3544.2$ kJ/kg

$h_6 = h_g - (1 - x_6)(h_g - h_f) = 2554.5 - 0.1979 \times 2433.1 = 2073.0$ kJ/kg

$Q_{out} = (h_6 - h_1) = 2073.0 - 121.4 = 1951.6$ kJ/kg

$W_{net} = \left(\text{Total } Q_{in}\right) - Q_{out} = 3544.2 - 1951.6 = 1592.6$ kJ/kg

∴ Cycle efficiency, $\eta_{CY} = \frac{1592.6}{3544.2} \times 100 = \underline{44.94\ \%}$

(2b) and (2c) With reheat pressures of 4 MN/m² and 2 MN/m²

The calculations proceed exactly as above, with the following results:

Reheat pressure, p_R	MN/m^2	4	2
h_5	kJ/kg	3445.0	3467.3
$s_6 = s_5$	kJ/kg K	7.0909	7.4323
$(1 - x_6)$		0.1719	0.1295
Exhaust wetness	%	*17.2*	*13.0*
t_4	°C	297.0	214.35
h_4	kJ/kg	2953.2	2803.1
h_5	kJ/kg	3445.0	3467.3
In reheater, Q_{in}	kJ/kg	491.8	664.2
In boiler proper, Q_{in}	kJ/kg	3174.1	3174.1
Total Q_{in}	kJ/kg	3665.9	3838.3
h_6	kJ/kg	2136.2	2239.3
Q_{out}	kJ/kg	2014.8	2117.9
W_{net}	kJ/kg	1651.1	1720.4
Cycle efficiency, η_{CY}	%	*45.04*	*44.82*

When these values of η_{CY} are plotted against p_R (with, preferably, the results from further calculations intermediate between these values of p_R), we find:

Optimum reheat pressure ≈ 3.75 MN/m^2

and
$$\frac{\text{optimum reheat pressure}}{\text{initial pressure}} \approx \frac{3.75}{15} = \frac{1}{4}$$

At the optimum reheat pressure, $\eta_{CY} = 45.04$ %

But $\eta_{RANK} = 43.59$ %

Improvement in efficiency due to reheating $= \frac{1.45}{43.59} \times 100 = 3.3$ %

Check calculation at $p_R = 4$ MN/m^2, using equation (7.18)

Referring to the hypothetical "cycles" N and R of Fig. 7.10 and Fig. P.7.11:

$$\left(\frac{\eta_{CY}}{\eta_N} - 1\right) = \frac{\left(\dfrac{\eta_R}{\eta_N} - 1\right)}{\left(\dfrac{Q_N}{Q_R} + 1\right)}. \tag{7.18}$$

<u>From the calculations for $p_R = 4\ MN/m^2$:</u>

<u>For "cycle" R:</u>

$Q_{in} = (h_5 - h_4) = 3445.0 - 2953.2 = 491.8$ kJ/kg

$Q_{out} = (h_6 - h_7) = 2136.2 - 1911.9 = \underline{224.3}$ kJ/kg

$W_{net} = 267.5$ kJ/kg

$\eta_R = \dfrac{267.5}{491.8} \times 100 = 54.39$ %

<u>For "cycle" N:</u> $\eta_N = \eta_{RANK} = 43.59$ %

Hence $\left(\dfrac{\eta_R}{\eta_N} - 1\right) = \left(\dfrac{54.39}{43.59} - 1\right) = 0.2478$

and $\left(\dfrac{Q_N}{Q_R} + 1\right) = \left(\dfrac{3174.1}{491.8} + 1\right) = 7.454$

∴ From eqn. (7.18), $\left(\dfrac{\eta_{CY}}{\eta_N} - 1\right) = \dfrac{0.2478}{7.454} \times 100 = \underline{3.3}$ % (as before)

<u>7.12</u> A steam power plant incorporates a homogeneous train of surface feed heaters with the bled-steam drains cascaded. In each heater the outlet feed water is raised to the saturation temperature of the entering bled steam, and the outlet drain water is cooled to the temperature of the entering feed water. Assuming that the enthalpy of water at any temperature is equal to the enthalpy of saturated water at the same temperature and that stray heat losses are negligible, derive an expression for the ratio of the mass flow rates of steam in the turbine upstream and downstream of a bled-steam tapping point.

A multi-reheat plant incorporating such a train of surface feed heaters is to be designed for a specified vacuum and final feed temperature, and for specified steam conditions in the boiler and reheaters. It may be assumed that β, the difference between the specific enthalpies of the bled steam entering a heater and of saturated water at the same pressure, will be the same for all bled-steam tapping points between any two adjacent reheat points, and that the condition line of the steam in the turbine will be unaffected by any changes in the positions of the bled-steam tapping points.

Show that, for maximum cycle efficiency, the enthalpy rises r of the feed water in all heaters between adjacent reheat points must be the same, and that the enthalpy rises in the sets of heaters taking steam from the turbine at points respectively immediately upstream and downstream of a selected reheat point must be such as to satisfy the relation

$$\frac{\alpha_d}{\alpha_u} = 1 + \frac{Q_S}{Q_B},$$

where $\alpha \equiv \beta + r$ and the suffixes u and d refer to conditions immediately upstream and downstream of the reheater respectively, Q_S is the heat supplied to the steam in the selected reheater and Q_B is the heat supplied in the boiler together with all reheaters upstream of the selected reheater.

Solution

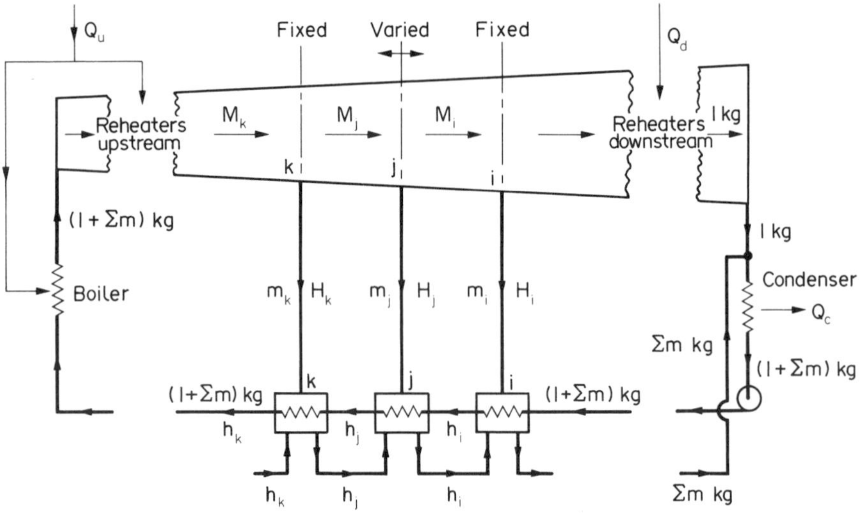

Fig. P.7.12(a).

Let $r_j \equiv (h_j - h_i) \equiv$ enthalpy rise of feed water in heater j,

and $R \equiv (h_k - h_i) \equiv$ total enthalpy rise (constant) of feed water in heaters j and k,

then $r_k = (R - r_j) \equiv$ enthalpy rise of feed water in heater k,

where heaters i, j and k are <u>any</u> three feed heaters <u>between</u> adjacent reheat points.

For <u>any</u> heater in which the enthalpy rise is r, the enthalpies will be as depicted in Fig. P.7.12(b).

Since the bled-steam drains are cascaded, as depicted in Fig. P.7.12(a), there will be $(1 + \Sigma m)$ kg of feed water through all the heaters (and passing to the boiler), <u>per kg</u>

of steam flowing to the condenser at turbine exhaust, where Σm is the sum of the bled steam quantities supplied to all the feed heaters in the system. The mass flows and specific enthalpies in the turbine and feed system adjacent to the given bled steam point will consequently be as depicted in Fig. P.7.12(b).

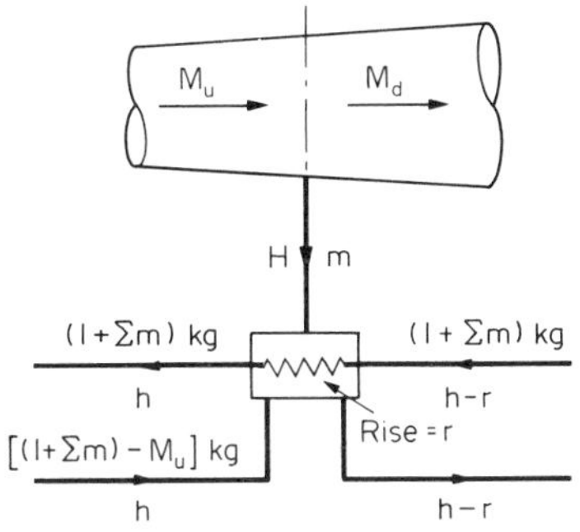

Fig. P.7.12(b).

For the feed heater depicted in this figure, the steady-flow energy equation gives:

$$[H - (h - r)]m + [(1 + \Sigma m) - M_u]r = (1 + \Sigma m)r$$

$$m(\beta + r) = M_u r$$

$$\therefore\ m = \frac{r}{\beta + r} M_u, \qquad (1)$$

where M_u is the steam flow rate in the turbine upstream of the given heater.

$$\text{Now } M_d = M_u - m$$

$$\therefore\ M_d = \frac{\beta}{\beta + r} M_u, \qquad (2)$$

where M_d is the steam flow rate in the turbine downstream of this heater.

Hence, for all feed heaters between two adjacent reheat points, we have the following general relation:

$$\boxed{\frac{M_u}{M_d} = \gamma}, \qquad (3)$$

$$\text{where } \gamma \equiv \frac{\beta + r}{\beta} \equiv \frac{\alpha}{\beta} \qquad (4)$$

$$\text{and } \alpha \equiv \beta + r. \qquad (5)$$

We may now turn to the problem of demonstrating that, if β is assumed to be the same for all feed heaters between any two adjacent reheat points, the cycle efficiency η_{CY} will be a maximum when the enthalpy rises are the same in all of these heaters.

The cycle efficiency, η_{CY}, is given by

$$\eta_{CY} = 1 - \frac{Q_c}{Q_u + Q_d} \tag{6}$$

where Q_u ≡ heat supplied in the boiler and in all reheaters upstream of the turbine section depicted in Fig. P.7.12(a),

Q_d ≡ heat supplied in all reheaters downstream of that section,

Q_c ≡ heat rejected in the condenser.

Keeping all bled-steam tapping points fixed except for that to heater j, we consider as our chosen independent variable r_j, the enthalpy rise of the feed water across heater j; while the total enthalpy rise across heaters j and k together will remain constant at R. Thus the enthalpy rise across heater k will also vary as we take different positions of tapping point j, and will be equal to $(R - r_j)$.

Per unit mass flow to the condenser, we know from the recurrence relation represented by equation (3) that the steam flows in all sections of the turbine upstream from the condenser up to and including M_i will remain constant as we vary r_j in consequence of varying the position of bled steam tapping j. The quantities Q_d and Q_c will consequently also remain constant. Thus, from equation (6), η_{CY} will be a maximum when Q_u is a maximum. Moreover, because none of the bled-steam tapping points upstream of tapping k are altered, we know from the recurrence relation (3) that Q_u will be proportional to M_k.

Thus $Q_u = k\, M_k$, where k is a constant. (7)

Applying equation (3) to tapping points i and j:

$$\frac{M_k}{M_i} = \frac{M_k}{M_j} \cdot \frac{M_j}{M_i} = \gamma_j\, \gamma_k. \tag{8}$$

Thus, since M_i is constant, Q_u, and thereføre η_{CY}, will be a maximum when $\gamma_j\, \gamma_k$ is a maximum. In terms of our chosen variable, r_j:

$$\gamma_j\, \gamma_k = \left(1 + \frac{r_j}{\beta}\right)\left(1 + \frac{R - r_j}{\beta}\right),$$

and $\underline{\eta_{CY}}$ will be a maximum when

$$\frac{d(\gamma_j \gamma_k)}{dr_j} = 0.$$

Hence

$$-\frac{1}{\beta}\left(1 + \frac{r_j}{\beta}\right) + \frac{1}{\beta}\left(1 + \frac{R - r_j}{\beta}\right) = 0$$

$$\therefore r_j = \frac{R}{2},$$

$$\text{so that } \underline{r_k = r_j} \qquad (9)$$

But these are any two heaters between adjacent reheaters. Hence the enthalpy rises must be the same in all feed heaters between adjacent reheaters.

Finally, we turn to the problem of determining that relation between the enthalpy rises in heaters respectively upstream and downstream of a selected reheater which will lead to maximum η_{CY}. For this purpose, we consider the situation depicted in Fig. P.7.12(c).

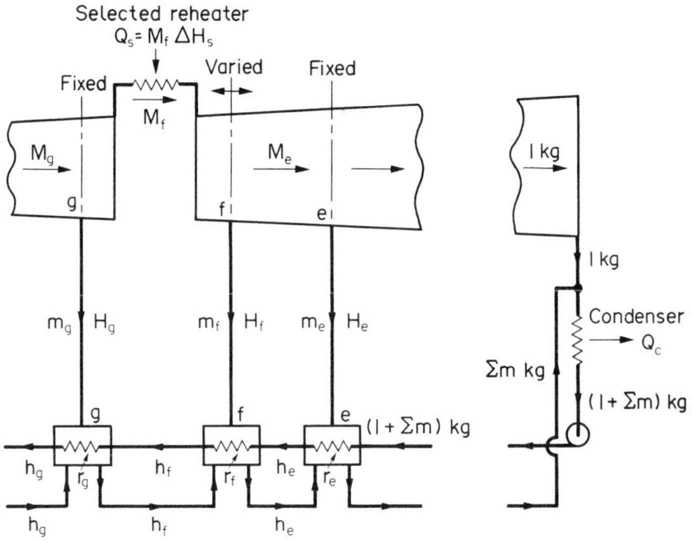

Fig. P.7.12(c).

For the purposes of the present analysis, the enthalpy rise r_f across heater f will be taken as our chosen independent variable. r_f is varied by varying the position of the bled-steam tapping point for heater f, whilst the tapping points for heaters e and g, and for all other heaters, remain unchanged. As r_f is varied, the tapping point for heater f will be assumed to remain downstream of the selected reheater; this is no real restriction, since it merely means that heaters f and g are those two heaters which, after optimization, will be found to straddle this reheater.

The cycle efficiency is given by

$$\eta_{CY} = 1 - \frac{Q_C}{Q_B + Q_S + Q_D}, \qquad (10)$$

where $Q_B \equiv$ heat supplied in the boiler and in all reheaters upstream of the selected reheater,

$Q_S \equiv$ heat supplied in the selected reheater,

$Q_D \equiv$ heat supplied in all reheaters downstream of the selected reheater,

$Q_C \equiv$ heat rejected in the condenser.

Per unit mass flow to the condenser, we know from the recurrence relation represented by equation (3) that the steam flows in all sections of the turbine upstream from the condenser up to and including M_e will remain constant as we vary r_f in consequence of varying the position of bled steam tapping f. The quantities Q_D and Q_C will consequently also remain constant. Moreover, because none of the bled-steam tapping points upstream of tapping g are altered, we know from the recurrence relation (3) that Q_B will be proportional to M_g.

Thus $Q_B = \mu M_g$, where μ is a constant.

Applying equation (3) to tapping points f and g:

$$\frac{M_g}{M_e} = \frac{M_g}{M_f} \cdot \frac{M_f}{M_e} = \gamma_g \gamma_f, \qquad (11)$$

where

$$\gamma_f \equiv \left(1 + \frac{r_f}{\beta_f}\right), \qquad (12)$$

$$\gamma_g \equiv \left(1 + \frac{R - r_f}{\beta_g}\right) \qquad (13)$$

and R is the constant total enthalpy rise across heaters f and g.

$$\text{Now } Q_B = \mu M_g = \mu \gamma_g \gamma_f M_e \qquad (14)$$

$$\text{and } Q_S = \Delta H_S M_f = \Delta H_S \gamma_f M_e, \qquad (15)$$

where ΔH_S is the enthalpy rise of the steam passing through the selected reheater.

Since Q_C and Q_D are constant, η_{CY} will be a maximum when $(Q_B + Q_S)$ is a maximum. Since r_f is our chosen independent variable, η_{CY} will thus be a maximum when

$$\frac{d(Q_B + Q_S)}{dr_f} = 0. \tag{16}$$

From equations (12) to (15):

$$\frac{dQ_B}{dr_f} = \mu M_e \left(\frac{\gamma_g}{\beta_f} - \frac{\gamma_f}{\beta_g}\right) = \frac{Q_B}{\gamma_g \gamma_f}\left(\frac{\gamma_g}{\beta_f} - \frac{\gamma_f}{\beta_g}\right) = Q_B\left(\frac{1}{\alpha_f} - \frac{1}{\alpha_g}\right), \tag{17}$$

$$\frac{dQ_S}{dr_f} = \frac{\Delta H_S M_e}{\beta_f} = \frac{Q_S}{\gamma_f \beta_f} = \frac{Q_S}{\alpha_f}, \tag{18}$$

where, from equation (4), $\alpha = \beta\gamma$.

Thus, from equations (16), (17) and (18), η_{CY} will be a maximum when

$$Q_B\left(\frac{1}{\alpha_f} - \frac{1}{\alpha_g}\right) + \frac{Q_S}{\alpha_f} = 0.$$

Thus, for maximum cycle efficiency, the bled-steam tapping point for feed heater f must be such as to satisfy the relation

$$\frac{\alpha_f}{\alpha_g} = 1 + \frac{Q_S}{Q_B}. \tag{19}$$

Now the reheater was any selected reheater, while f and g relate to the bled-steam tapping points respectively downstream and upstream of this reheater. Thus, for any reheater in general, we shall have the required relation

$$\frac{\alpha_d}{\alpha_u} = 1 + \frac{Q_S}{Q_B},$$

where $\alpha \equiv \beta + r$.

Note: The derivation of a relation similar to this for a system of direct-contact feed heaters appears in Reference 15 of Chapter 7, namely: C. D. Weir, Optimization of Heater Enthalpy Rises in Feed-Heating Trains; written communication by R. W. Haywood. That analysis did not assume constancy of β between adjacent reheat points.

Note: Problem 8.6 in Chapter 8, relating to a steam cycle for a nuclear power plant, provides further practice in feed-heating calculations, as also do Problems 9.2 and 9.4 in Chapter 9.

7.13 Figure 7.11 depicts a plant in which a double-pass-out steam turbine is used for the combined supply of power and process steam. All the process steam is returned as condensate and this is mixed with the condensate from the condenser before the joint flow enters the deaerator feed heater. The latter is supplied with steam from the second pass-out point and delivers saturated water at the temperature indicated in the figure.

For the conditions depicted in the figure, and taking values of 85 % and 97 % respectively for the overall isentropic efficiency of the turbine and the efficiency of the generator, calculate the following quantities:

(1) The mass flow rates $\dot{m}_C$ and $\dot{m}_H$.

(2) The electrical output from the plant.

(3) The energy given up by the process steam.

(4) The work efficiency and total efficiency of the plant.

Enthalpy rises in the pumps are to be neglected.

Solution

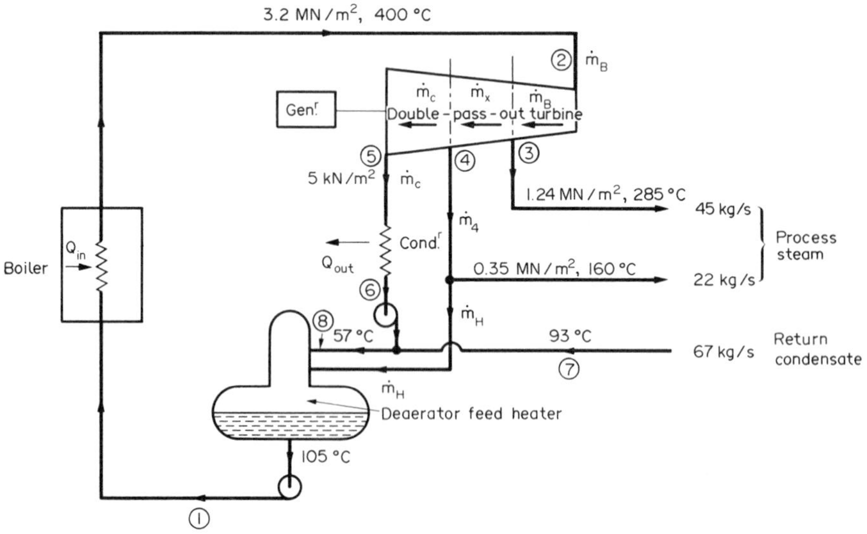

Fig. P.7.13 (Similar to Fig. 7.11).

(1) Using the notation of Fig. P.7.13, let 5_s denote the state of steam after isentropic expansion from the turbine-inlet condition to turbine-exhaust pressure.

<u>Expansion in turbine</u>:

At 3.2 MN/m² and 400 °C:

$h_2 = 3229$ kJ/kg, $s_2 = 6.891$ kJ/kg K

At 5 kN/m²:

$s_g = 8.396$ kJ/kg K, $s_f = 0.476$ kJ/kg K

$h_g = 2561.6$ kJ/kg, $h_f = 137.8$ kJ/kg

Dryness fraction, $x5_s$, at state 5_s:

$$(1 - x5_s) = \frac{8.396 - 6.891}{8.396 - 0.476} = \frac{1.505}{7.920} = 0.1900$$

$$h_{5_s} = 2561.6 - 0.1900 \times 2423.8 = 2101 \text{ kJ/kg}$$

$$\Delta h_s \equiv (h_2 - h5_s) = 1128 \text{ kJ/kg}$$

$$\Delta H \equiv \eta_T \, \Delta H_s = 0.85 \times 1128 = 959 \text{ kJ/kg}$$

$$h_5 = (h_2 - \Delta H) = \underline{2270} \text{ kJ/kg}$$

Specific enthalpies

At 1.24 MN/m² and 285 °C, $h_3 = 3013$ kJ/kg

At 0.35 MN/m² and 160 °C, $h_4 = 2778$ kJ/kg

At 5 kN/m², $h_6 = h_f = 137.8$ kJ/kg

At 93 °C, $h_7 = h_f = 389.6$ kJ/kg

At 57 °C, $h_8 = h_f = 238.5$ kJ/kg

At 105 °C, $h_1 = h_f = 440.2$ kJ/kg

Mixing of return condensate and condensate from condenser:

$$\dot{m}_C h_6 + 67\, h_7 = (\dot{m}_C + 67) h_8$$

$$\dot{m}_C \times 137.8 + 67 \times 389.6 = (\dot{m}_C + 67) \times 238.5$$

$$\dot{m}_C = \frac{67 \times 151.1}{100.7} = \underline{100.53} \text{ kg/s}$$

$$\therefore \dot{m}_8 = 100.53 + 67 = 167.53 \text{ kg/s}$$

Steady-flow energy equation for the deaerator feed heater:

$$\dot{m}_8 h_8 + \dot{m}_H h_4 = (\dot{m}_8 + \dot{m}_H) h_1$$

$$167.53 \times 238.5 + 2778\, \dot{m}_H = (167.53 + \dot{m}_H) \times 440.2$$

$$\dot{m}_H = \frac{167.53 \times 201.7}{2337.8} = \underline{14.45} \text{ kg/s}$$

Other flow rates

$\dot{m}_4 = \dot{m}_H + 22 = 36.45$ kg/s

$\dot{m}_x = \dot{m}_4 + \dot{m}_C = 136.98$ kg/s

$$\dot{m}_B = \dot{m}_x + 45 = 181.98 \text{ kg/s}$$

(2) Turbine power output and electrical output

$$\dot{m}_B(h_2 - h_3) = 181.98 \times 216 \times 10^{-3} = 39.31 \text{ MW}$$

$$\dot{m}_x(h_3 - h_4) = 136.98 \times 235 \times 10^{-3} = 32.19 \text{ MW}$$

$$\dot{m}_C(h_4 - h_5) = 100.53 \times 508 \times 10^{-3} = 51.07 \text{ MW}$$

Total turbine power output = 122.57 MW

∴ Electrical output = 122.57 x 0.97 = 118.9 MW

(3) Thermal output (energy given up by process steam)

$$45(h_3 - h_7) = 45 \times 2623.4 \times 10^{-3} = 118.05 \text{ MW}$$

$$22(h_4 - h_7) = 22 \times 2388.4 \times 10^{-3} = 52.54 \text{ MW}$$

Total thermal output = 170.6 MW

(4) Plant efficiencies

$$\dot{Q}_{in} = \dot{m}_B(h_2 - h_1) = 181.98 \times 2788.8 \times 10^{-3} = 507.5 \text{ MW}$$

Neglecting pump work inputs:

$$\text{Work efficiency} \equiv \frac{\dot{W}_{net}}{\dot{Q}_{in}} = \frac{118.9}{507.5} \times 100 = 23.4\ \%$$

$$\text{Total efficiency} \equiv \frac{W_{net} + \text{Energy to process plant}}{Q_{in}}$$

$$\therefore \text{Total efficiency} = \left(\frac{118.9 + 170.6}{507.5}\right) \times 100 = \frac{289.5}{507.5} \times 100 = 57.0\ \%$$

Note: For further study of this plant, see J. C. Robertson, Power plant energy conservation, Proc. Amer. Pwr. Conf., Vol. 37, p. 671 (1975).

CHAPTER 8

Nuclear Power Plant

For the solution of Problems 8.1 to 8.6, use has been made of the UK Steam Tables in SI Units 1970, Ed. Arnold (Publishers) Ltd., London (1970).

Problems 8.1 to 8.5 relate to the Calder Hall (gas-cooled) type of plant illustrated in Figs. 8.1 and 8.2.

8.1 At the design load the feed water is heated just to the saturation temperature of the LP steam in each section of the mixed economiser, the minimum temperature approach between the two fluids is 17 K in both the HP and LP heat exchangers, and the conditions in the steam-raising towers are as follows:

Temperature of entering CO_2	= 337 °C
Temperature of entering feed water	= 38 °C
HP steam conditions at exit	= 1.45 MN/m², 316 °C
LP steam conditions at exit	= 0.435 MN/m², 177 °C

Treating CO_2 as a perfect gas with c_p = 1.017 kJ/kg K, and neglecting pressure drops and external heat losses in the heat exchangers, make the following calculations for the plant per kg of CO_2 circulated:

(1) Calculate the masses of HP and LP steam produced, the temperature of the CO_2 at exit from the towers, and the heat transferred in the steam generators.

(2) Taking as the environment temperature that of the circulating water at inlet to the condenser, which is at 24 °C, calculate the energy available for the production of work from (a) the heat transferred from the CO_2 in its passage through the steam generators, and (b) the H_2O. In each case express the available energy as a percentage of the heat transferred in the steam generators.

(3) Calculate the reduction in the available energy of the steam

if the environment temperature is taken as being the saturation temperature of the steam in the condenser, in which the pressure is 6 kN/m². Express the new available energy as a percentage of the heat transferred in the steam generators.

(4) If the dual-pressure cycle were to be replaced by a single-pressure cycle, producing steam at 316 °C from feed water at 38 °C, with the inlet and exit temperatures of the CO_2 and the minimum temperature approach between the two fluids the same as in the dual-pressure cycle, determine (a) the steam pressure, and the mass of steam produced, (b) the available energy for an environment temperature of 24 °C, again expressing it as a percentage of the heat transferred in the steam generators.

Solution

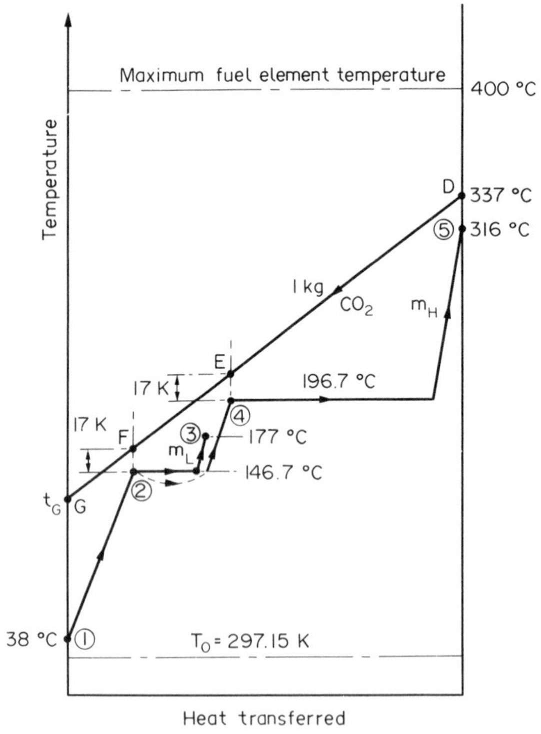

Fig. P.8.1(a)

At 1.45 MN/m² (14.5 bar):

Saturation temperature = 196.7 °C

At 0.435 MN/m^2 (4.35 bar):

Saturation temperature = 146.7 °C

(1) Applying the steady-flow energy equation to the respective sections of the heat exchangers:

<u>D to E</u>

$$m_H(h_5 - h_4) = c_p(t_D - t_E)$$

$$m_H(3075.6 - 837.5) = 1.017\ (337 - 213.7)$$

$$m_H = \frac{1.017 \times 123.3}{2238.1} = \underline{0.0560} \text{ kg}$$

<u>E to F</u>

$$m_L(h_3 - h_2) + m_H(h_4 - h_2) = c_p(t_E - t_F)$$

$$m_L(2809.0 - 617.8) + 0.0560(837.5 - 617.8) = 1.017(213.7 - 163.7)$$

$$m_L = \frac{50.85 - 12.30}{2191.2} = \underline{0.0176} \text{ kg}$$

<u>F - G</u>

$$(m_L + m_H)(h_2 - h_1) = c_p(t_F - t_G)$$

$$0.0736(617.8 - 159.1) = 1.017(163.7 - t_G)$$

$$(163.7 - t_G) = \frac{33.76}{1.017} = 33.2 \quad \therefore\ t_G = \underline{130.5}\ °C$$

<u>Heat transferred</u> $= c_p(t_D - t_G) = 1.017(337 - 130.5)$

$= \underline{210.0}$ kJ per kg of CO_2 circulated

(2) <u>Energy available for the production of work</u>

(a) <u>From the heat transferred from the CO_2</u>:

$$dW = \left(1 - \frac{T_0}{T}\right)(-c_p\ dT)$$

<u>Available energy</u> $= - c_p \int_D^G \left(1 - \frac{T_0}{T}\right) dT$

$$= c_p(T_D - T_G) - c_p\ T_0\ \ln \frac{T_D}{T_G}$$

$$= 210.0 - 1.017 \times 297.15 \ln \frac{610.15}{403.65}$$

$$= 210.0 - 124.9 = \underline{85.1} \text{ kJ}$$

<u>Percentage</u> (of heat transferred) $= \frac{85.1}{210.0} \times 100 = \underline{40.5}$ %

(b) From the H_2O:

High-pressure H_2O

$$\text{Available energy per kg} = (h_5 - h_1) - T_0(s_5 - s_1)$$
$$= (3075.6 - 159.1) - 297.15(6.9989 - 0.5453)$$
$$= 2916.5 - 1917.7 = 998.8 \text{ kJ/kg}$$
$$\text{Available energy} = 998.8 \times 0.0560 = 55.9 \text{ kJ}$$

Low-pressure H_2O

$$\text{Available energy per kg} = (h_3 - h_1) - T_0(s_3 - s_1)$$
$$= (2809.0 - 159.1) - 297.15(7.0217 - 0.5453)$$
$$= 2649.9 - 1924.5 = 725.4 \text{ kJ/kg}$$
$$\text{Available energy} = 725.4 \times 0.0176 = 12.8 \text{ kJ}$$

$\therefore$ Total available energy from H_2O = 55.9 + 12.8 = 68.7 kJ

Percentage (of heat transferred) $= \dfrac{68.7}{210.0} \times 100 = 32.7$ %

(3) Available energy, with T_0 equal to condenser saturation temperature

At 6 kN/m², saturation temperature = 36.2 °C

Increase in T_0 = 36.2 - 24.0 = 12.2 K

Decrease in available energy:

$$\text{High-pressure } H_2O = m_H(s_5 - s_1) \times 12.2$$
$$= 0.0560 \times 6.4536 \times 12.2 = 4.41 \text{ kJ}$$
$$\text{Low-pressure } H_2O = m_L(s_3 - s_1) \times 12.2$$
$$= 0.0176 \times 6.4764 \times 12.2 = 1.39 \text{ kJ}$$

Total decrease = 5.8 kJ

New available energy = 68.7 - 5.8 = 62.9 kJ

Percentage (of heat transferred) $= \dfrac{62.9}{210.0} \times 100 = 30.0$ %

(4) Single-pressure cycle

Applying the steady-flow energy equation to the respective sections of the heat exchanger:

A to B $m(h_7 - h_6) = c_p(t_A - t_B)$

A to C $m(h_7 - h_1) = c_p(t_A - t_C)$

$$\therefore \frac{h_7 - h_6}{h_7 - h_1} = \frac{337 - t_B}{206.5}$$

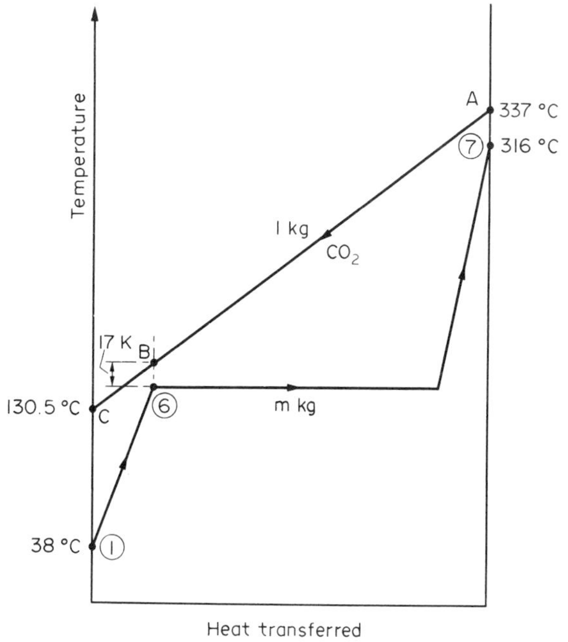

Fig. P.8.1(b)

(a) Calculation of the steam pressure and the mass of steam produced

Determined by iterative procedure, or alternatively (as below), checking that p = 0.418 MN/m² satisfies the above equation.

At p = 0.418 MN/m² (4.18 bar), saturation temperature = 145.2 °C

$\therefore t_B = 162.2$ °C

At 0.418 MN/m² and 316 °C, $h_7 = 3099.7$ kJ/kg; $h_6 = 611.5$ kJ/kg

Putting these values in the above equation:

$$\frac{3099.7 - 611.5}{3099.7 - 159.1} = \frac{337 - 162.2}{206.5}$$

$$0.8462 = 0.8465$$ Hence equation satisfied.

$\therefore$ Steam pressure = *0.418* MN/m²

Per kg of CO_2 circulated:

Mass of steam produced, $m = \frac{c_p(T_A - T_C)}{h_7 - h_1} = \frac{210.0}{2940.6} = \underline{0.0714}$ kg

(b) Calculation of the available energy from the H_2O

Available energy per kg $= (h_7 - h_1) - T_0(s_7 - s_1)$

$= 2940.6 - 297.15(7.6032 - 0.5453)$

$= 2940.6 - 2097.3 = 843.3$ kJ/kg

Available energy $= 843.3 \times 0.0714 = \underline{60.2}$ kJ/kg CO_2

Percentage (of heat transferred) $= \frac{60.2}{210.0} \times 100 = \underline{28.7}$ %

8.2 There is a pressure drop of 70 kN/m² and a temperature drop of 6 K in each of the steam mains between the steam generators and the turbine, and the condenser pressure is 6 kN/m². The internal isentropic efficiency of the HP section of the turbine is 85 % and of the LP section is 80 %.

(1) Determine the specific enthalpy of the steam at the end of the HP section of the turbine, and the shaft work obtained from this section per kg of CO_2 circulated through the reactor.

(2) Assuming perfect mixing of the HP exhaust steam and LP inlet steam, determine the specific enthalpy of the steam entering the first stage of the LP section of the turbine, the specific enthalpy of the steam at LP turbine exhaust, and the shaft work obtained from this section of the turbine per kg of CO_2 circulated through the reactor.

(3) The combined efficiency factor for the turbine external losses and alternator losses is 94 %. Calculate the output from the alternator terminals per kg of CO_2 circulated through the reactor, and express this output as a percentage of the heat transferred in the steam generators.

Solution

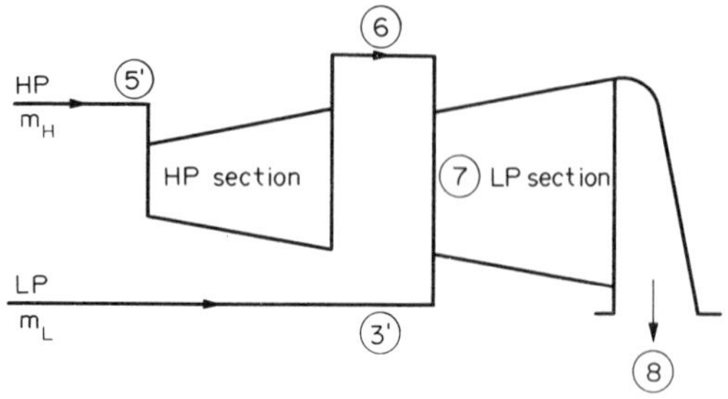

Fig. P.8.2.

(1) At HP turbine inlet:

Pressure = 1.45 - 0.07 = 1.38 MN/m^2

Temp. = 316 - 6 = 310 °C

$h_{5'}$ = 3064.2 kJ/kg

$s_{5'}$ = 7.0017 kJ/kg K

HP turbine

HP exhaust pressure = LP inlet pressure, $p_{3'}$ = 0.435 - 0.070 = 0.365 MN/m^2

For isentropic expansion at s = 7.0017 kJ/kg K to p = 0.365 MN/m^2:

At p = 0.360 MN/m^2, $h = 2755.4 + \frac{0.0181}{0.0511} \times 21.9$ = 2763.2 kJ/kg

At p = 0.370 MN/m^2, $h = 2754.6 + \frac{0.0324}{0.0514} \times 21.9$ = 2768.4 kJ/kg

∴At p = 0.365 MN/m^2, s = 7.0017 kJ/kg, h_{6_s} = 2765.8 kJ/kg

Isentropic enthalpy drop, Δh_s = 3064.2 - 2765.8 = 298.4 kJ/kg

∴ Actual enthalpy drop, Δh = 298.4 x 0.85 = 253.6 kJ/kg

At HP turbine exhaust, h_6 = 3064.2 - 253.6 = 2810.6 kJ/kg

HP turbine output = $m_H \Delta h$ = 0.0560 x 253.6 = 14.20 kJ/kg CO_2

(2) At LP turbine inlet:

Pressure = 0.365 MN/m^2

Temp. = 177 - 6 = 171 °C

$h_{3'}$ = 2800.7 kJ/kg

Mixing at LP inlet:

$$(m_H + m_L)h_7 = m_H h_6 + m_L h_{3'}$$

$$0.0736\ h_7 = 0.0560 \times 2810.6 + 0.0176 \times 2800.7$$

At LP turbine inlet, $h_7 = \frac{206.7}{0.0736}$ = 2808.4 kJ/kg

Calculation of s_7, when h_7 = 2808.4 kJ/kg:

At p = 0.360 MN/m^2, $s = 7.0840 + \frac{9.5}{21.4} \times 0.0477$ = 7.1052 kJ/kg K

At p = 0.370 MN/m^2, $s = 7.0701 + \frac{10.2}{21.4} \times 0.0479$ = 7.0929 kJ/kg K

∴ At p = 0.365 MN/m^2, s_7 = 7.0991 kJ/kg K

<u>LP turbine</u>

Exhaust pressure = 6 kN/m²

For isentropic expansion at s = 7.0991 kJ/kg K to p = 6 kN/m²:

Wetness fraction at exhaust, $(1 - x_{8_s}) = \frac{s_g - s_7}{s_{fg}} = \frac{8.3312 - 7.0991}{7.8103} = 0.1578$

$h_{8_s} = h_g - (1 - x_{8_s})h_{fg} = 2567.5 - 0.1578 \times 2416.0 = 2186.3$ kJ/kg

Isentropic enthalpy drop, $\Delta h_s = 2808.4 - 2186.3 = 622.1$ kJ/kg

∴ Actual enthalpy drop, $\Delta h = 622.1 \times 0.80 = 497.7$ kJ/kg

<u>At LP turbine exhaust</u>, $h_8 = 2808.4 - 497.7 = \underline{2310.7}$ kJ/kg

<u>LP turbine output</u> = $(m_H + m_L)\Delta h = 0.0736 \times 497.7 = \underline{36.63}$ kJ/kg CO_2

(3) Gross turbine output = 14.20 + 36.63 = 50.83 kJ/kg CO_2

∴ <u>Alternator output</u> = 50.83 x 0.94 = <u>*47.8*</u> kJ/kg CO_2

<u>Percentage</u> (of heat transferred) = $\frac{47.8}{210.0} \times 100 = \underline{22.8}$ %

8.3 The pressure of the CO_2 at inlet to the gas circulators is 0.8 MN/m²; they provide a pressure rise of 28 kN/m² and their internal isentropic efficiency is 80 %. Estimate the temperature rise of the CO_2 in its passage through the circulators, and calculate the thermal output of the reactor per kg of CO_2 circulated.

Solution

<u>In the gas circulators</u>:

Mean pressure = 800 + 14 = 814 kN/m²

Estimated mean temperature ≈ 132 °C = 405.15 K

Mean specific volume, $v = \frac{RT}{p} = \frac{189 \times 405.15}{814 \times 10^3} = 0.0941$ m³/kg

Isentropic enthalpy rise, $\Delta h_s = v\Delta p = 0.0941 \times 28 = 2.635$ kJ/kg

Actual enthalpy rise, $\Delta h = \frac{\Delta h_s}{\eta_c} = \frac{2.635}{0.80} = 3.294$ kJ/kg

<u>Actual temperature rise</u> = $\frac{\Delta h}{c_p} = \frac{3.294}{1.017} = \underline{3.2}$ K

[∴ Mean temperature = 130.5 + 1.6 = 132.1 °C]

Temperature at reactor inlet = 130.5 + 3.2 = 133.7 °C

Temperature at reactor outlet = 337 °C (given in Problem 8.1)

∴ Temperature rise in reactor = 203.3 K

∴ Reactor thermal output = $c_p \Delta T$ = 1.017 x 203.3 = *206.7* kJ/kg CO_2

8.4 The combined efficiency factor for the external losses in the circulators and the losses in their driving motors is 93 %. At the design load the mass flow rate of CO_2 is 880 kg/s. Calculate the following quantities at this load, all expressed in MW:

(1) The thermal output of the reactor.

(2) The electrical output at the alternator terminals.

(3) The total input to the circulator motors. Express this as a percentage of (a) the electrical output at the alternator terminals, and (b) the reactor thermal output.

(4) The net electrical output of the station, and the overall station efficiency, neglecting the power required for other auxiliary plant. For comparison, calculate the ideal thermal efficiency of a cyclic heat power plant working between a source temperature equal to the maximum permissible fuel-element temperature of 400 °C and a sink temperature equal to that of the environment at 24 °C.

Solution

(1) Reactor thermal output = $880 \times 206.7 \times 10^{-3}$ = *181.9* MW

(2) Alternator output = $880 \times 47.8 \times 10^{-3}$ = *42.1* MW

(3) Input to circulator motors $= \dfrac{\dot{m}_{CO_2}\,\Delta h}{\eta_M} = \dfrac{880 \times 3.294 \times 10^{-3}}{0.93}$ = *3.1* MW

(a) Percentage of alternator output $= \dfrac{3.1}{42.1} \times 100$ = *7.4* %

(b) Percentage of reactor output $= \dfrac{3.1}{181.9} \times 100$ = *1.7* %

(4) Net station output = 42.1 - 3.1 = *39.0* MW

Overall station efficiency $= \dfrac{39.0}{181.9} \times 100$ = *21.4* %

Ideal Carnot cycle efficiency $= \dfrac{376}{673.15} \times 100$ = *55.9* %

8.5 The reactor uses natural uranium fuel, each tonne (10^3 kg) of which contains 7 kg of the fissile U^{235}, of which half can be consumed before the fuel elements have to be withdrawn from the reactor. The U^{235} is consumed at the rate of 1.3 g per megawatt-day of reactor thermal output. What steady rate of fuel replacement would be required if the reactor operated continuously at the design load?

If the heat transferred in the steam generators were supplied in a boiler of 85 % efficiency and burning coal having a calorific value of 25.6 MJ/kg, what would be the rate of coal consumption at the design load?

Solution

Rate of consumption of U^{235} = 1.3 x 181.9 = 236.5 g/day

Since 10^3 kg of the natural uranium fuel contain 7 kg of U^{235}, of which half is consumed before fuel-rod replacement is required:

$$\underline{\text{Rate of fuel replacement}} = \frac{236.5}{3.5} = \underline{67.6} \text{ kg/day}$$

In the steam-raising towers, rate of heat transfer

= 880 x 210.0 kW

$\therefore$ Corresponding rate of coal consumption

$$= \frac{880 \times 210.0}{25.6 \times 10^3 \times 0.85} \times \frac{3600 \times 24}{10^3}$$

$$= \underline{734} \text{ tonne/day}$$

8.6 For the operating pressures and temperatures in the PWR plant illustrated in Fig. 8.10, perform the following calculations. Frictional pressure drops and stray heat losses are to be neglected, and the boiler may be assumed to deliver dry saturated steam at the pressure indicated.

(1) Determine the steam wetness at HP turbine exhaust, given that the isentropic efficiency of the HP turbine is 81 %.

(2) Determine the fraction of the wet steam mixture supplied to the separator that is drained off as saturated water to the drain tank, given that the steam wetness at separator exit is 1 %.

(3) Determine the steam wetness at LP turbine exhaust, given that the isentropic efficiency of the LP turbine is 81 %.

It may be assumed that the heating steam is just condensed in the reheater and in all six feed heaters, and that the feed water leaving each heater is raised to the saturation temperature of the bled steam supplied to the heater. The work input to all pumps may be neglected, the specific enthalpy of water at any temperature being taken as equal to the saturation specific enthalpy at that temperature.

(4) From energy balances on the reheater, on feed heater 6 and on feed heater 5, calculate the values of the following ratios of the flow rates indicated in Fig. 8.10: (a) m_H/m_B, (b) m_R/m_B, (c) m_L/m_B.

The bled steam supplied to feed heater 6 is at a pressure of 2.55 MN/m² and its specific enthalpy is 2715 kJ/kg.

(5) From energy balances on feed heaters 4, 3, 2 and 1 in turn,

calculate the bled steam quantities supplied to these heaters, each expressed as a fraction of m_B, and thence evaluate m_D/m_B and m_C/m_B.

The following table gives the condition of the bled steam supplied to each of these feed heaters:

Heater No.	4	3	2	1
Pressure MN/m²	0.585	0.248	0.088	0.0245
Specific enthalpy, kJ/kg	2803	2674	2538	2392

(6) Determine the thermal efficiency of the steam cycle.

(7) From an energy balance for the boiler, determine the value of m_A/m_B.

(8) Taking the environment temperature as being equal to the saturation temperature of the steam in the condenser, calculate the energy available for the production of work from (a) the high-pressure fluid passing through the reactor, and (b) the low-pressure fluid passing through the boiler. In each case, express the available energy as a percentage of the thermal output of the reactor.

(9) Neglecting pressure drops on both sides of the boiler, determine the minimum temperature approach between the two fluids in their passage through the boiler (c.f. the pinch points discussed in Section 8.2 for gas-cooled reactors).

Solution

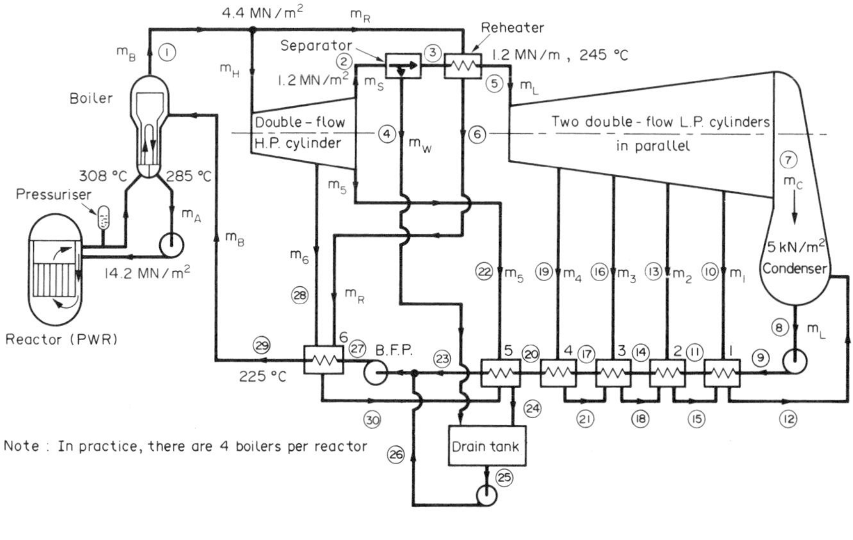

Fig. P.8.6(a). (Similar to Fig. 8.10)

Using the notation of Fig. P.8.6(a)

(1) HP turbine

At HP turbine inlet (4.4 MN/m², dry saturated):

$h_1 = 2798.3$ kJ/kg $\qquad s_1 = 6.0286$ kJ/kg K

At HP turbine exhaust (1.2 MN/m²)

Dryness fraction after isentropic expansion from state $1 \equiv x_{2_s}$

$$(1 - x_{2_s}) = \frac{s_g - s_1}{s_{fg}} = \frac{6.5194 - 6.0286}{4.3034} = 0.1140$$

$$h_{2_s} = 2782.7 - 0.1140 \times 1984.3 = 2556.5 \text{ kJ/kg}$$

Isentropic enthalpy drop, $\Delta h_s = 2798.3 - 2556.5 = 241.8$ kJ/kg

Actual enthalpy drop, $\Delta h = 241.8 \times 0.81 = 195.9$ kJ/kg

$h_2 = 2798.3 - 195.9 = 2602.4$ kJ/kg

Wetness at HP exhaust $= \dfrac{2782.7 - 2602.4}{1984.3} \times 100 = 9.1\ \%$

(2) Separator

$p_3 = p_2 = 1.2$ MN/m² $\qquad (1 - x_3) = 0.01$ (given)

$\therefore h_3 = 2782.7 - 0.01 \times 1984.3 = 2762.9$ kJ/kg

The steady-flow energy equation and the mass conservation equation respectively give:

$$m_S h_2 = m_L h_3 + m_W h_4$$

$$m_S = m_L + m_W$$

Whence fraction drained off is given by:

$$\frac{m_W}{m_S} = \frac{h_3 - h_2}{h_3 - h_4} = \frac{2762.9 - 2602.4}{2762.9 - 798.4} = 0.0817 \qquad \text{(a)}$$

(3) LP turbine

At LP turbine inlet (1.2 MN/m², 245 °C):

$h_5 = 2923.8$ kJ/kg $\qquad s_5 = 6.8082$ kJ/kg K

At LP turbine exhaust (5 kN/m²)

Dryness fraction after isentropic expansion from state $5 \equiv x_{7_s}$

$$(1 - x_{7_s}) = \frac{s_g - s_5}{s_{fg}} = \frac{8.3960 - 6.8082}{7.9197} = 0.2005$$

$$h_{7_s} = 2561.6 - 0.2005 \times 2423.8 = 2075.6 \text{ kJ/kg}$$

Isentropic enthalpy drop, $\Delta h_s = 2923.8 - 2075.6 = 848.2$ kJ/kg

Actual enthalpy drop, $\Delta h = 848.2 \times 0.81 = 687.0$ kJ/kg

$h_7 = 2923.8 - 687.0 = 2236.8$ kJ/kg

Wetness at LP exhaust $= \dfrac{2561.6 - 2236.8}{2423.8} \times 100 = 13.4$ %

(4) Flow rates to the turbines and the reheater

For the reheater:

$p_6 = p_1 = 4.4$ MN/m² $\qquad h_6 = 1115.4$ kJ/kg

Energy balance for the reheater gives:

$$m_R(h_1 - h_6) = m_L(h_5 - h_3)$$

Also $m_L = m_S - m_W$, so that, from eqn. (a),

$$m_L = 0.9183\, m_S \qquad \text{(b)}$$

$$m_R = \left(\frac{h_5 - h_3}{h_1 - h_6}\right)\left(\frac{m_L}{m_S}\right)m_S = \left(\frac{2923.8 - 2762.9}{2798.3 - 1115.4}\right) \times 0.9183\, m_S$$

$$\therefore\ m_R = 0.08780\, m_S \qquad \text{(c)}$$

For feed heater 6:

$p_{28} = 2.55$ MN/m² $\qquad h_{28} = 2715$ kJ/kg (given)

$h_{29} \approx h_{30} = (h_f$ at 2.55 MN/m²$) = 966.8$ kJ/kg

$h_{27} \approx h_{26} \approx h_{25} = h_{24}$

$h_{23} \approx h_{24} = h_4 = (h_f$ at 1.2 MN/m²$) = 798.4$ kJ/kg

Energy balance for feed heater 6 gives:

$$m_6(h_{28} - h_{30}) + m_R(h_6 - h_{30}) = m_B(h_{29} - h_{27})$$

$$m_6(2715 - 966.8) + m_R(1115.4 - 966.8) = m_B(966.8 - 798.4)$$

Thence, substituting for m_R from equation (c),

$$\underline{m_6 = 0.09633\ m_B - 0.00746\ m_S} \qquad (d)$$

Also $m_{30} = m_6 + m_R$

$$\therefore\ \underline{m_{30} = 0.09633\ m_B + 0.08034\ m_S} \qquad (e)$$

<u>For feed heater 5</u>:

From feed heater 4, $h_{20} \approx h_{21} = (h_f \text{ at } 0.585\ \text{MN/m}^2) = 666.2\ \text{kJ/k}$

Energy balance for feed heater 5 gives:

$$m_5(h_{22} - h_{24}) + m_{30}(h_{30} - h_{24}) = m_L(h_{23} - h_{20})$$

But $h_{22} = h_2 = 2602.4$ kJ/kg

$$\therefore\ m_5(2602.4 - 798.4) + m_{30}(966.8 - 798.4) = m_L(798.4 - 666.2)$$

$$\therefore\ m_5 = 0.07328\ m_L - 0.09335\ m_{30}$$

Thence, substituting for m_L from eqn. (b), and for m_{30} from eqn. (e),

$$\underline{m_5 = 0.05979\ m_S - 0.00899\ m_B} \qquad (f)$$

<u>For the HP turbine</u>:

$m_H = m_S + m_5 + m_6$, whence, from eqns. (d) and (f)

$$\underline{m_H = 1.05233\ m_S + 0.08734\ m_B} \qquad (g)$$

<u>For the boiler</u>:

$m_B = m_H + m_R$, whence, from eqns. (c) and (g),

$$m_B = (1.05233\ m_S + 0.08734\ m_B) + 0.08780\ m_S \qquad (h)$$

$$\text{Whence } m_S = \frac{0.91266}{1.14013}\ m_B = 0.80049\ m_B \qquad (i)$$

<u>Required flow ratios</u>:

From eqns. (g) and (i):

$$m_H = 1.05233 \times 0.80049\ m_B + 0.08734\ m_B$$

(a) $\therefore\ m_H/m_B = 0.9297 \approx \underline{0.930}$ (j)

From eqns. (c) and (i):

$m_R = 0.08780 \times 0.80049\ m_B$

(b) $\therefore\ m_R/m_B = 0.0703 \approx \underline{0.070}$ (k)

From eqns. (b) and (i):

$m_L = 0.9183 \times 0.80049\ m_B$

(c) $\therefore\ m_L/m_B = 0.7351 \approx \underline{0.735}$ (l)

(5) Bled steam quantities to feed heaters 1 to 4

Feed heater 4

$h_{20} \approx h_{21} = (h_f \text{ at } 0.585\ \text{MN/m}^2) = 666.2\ \text{kJ/kg}$

$h_{17} \approx h_{18} = (h_f \text{ at } 0.248\ \text{MN/m}^2) = 534.2\ \text{kJ/kg}$

$h_{19} = 2803\ \text{kJ/kg (given)}$

$m_4(h_{19} - h_{21}) = m_L(h_{20} - h_{17})$

$$\therefore\ m_4 = \frac{666.2 - 534.2}{2803 - 666.2}\ m_L = 0.06177\ m_L$$

Whence, from eqn. (l),

$$\frac{m_4}{m_B} = 0.06177 \times 0.7351 = \underline{0.04541} \qquad \text{(m)}$$

Feed heater 3

$h_{17} \approx h_{18} = 534.2\ \text{kJ/kg}$ $\qquad h_{16} = 2674\ \text{kJ/kg (given)}$

$h_{14} \approx h_{15} = (h_f \text{ at } 0.088\ \text{MN/m}^2) = 402.6\ \text{kJ/kg}$

$m_3(h_{16} - h_{18}) + m_4(h_{21} - h_{18}) = m_L(h_{17} - h_{14})$

Whence, from eqns. (l) and (m),

$2139.8\ m_3 + 0.04541 \times 132.0\ m_B = 0.7351 \times 131.6\ m_B$

$$\therefore\ \frac{m_3}{m_B} = \frac{90.75}{2139.8} = \underline{0.04241}$$

Feed Heater 2

$h_{14} \approx h_{15} = 402.6$ kJ/kg $h_{13} = 2538$ kJ/kg (given)

$h_{11} \approx h_{12} = (h_f$ at 0.0245 MN/m²$) = 270.1$ kJ/kg

$m_2(h_{13} - h_{15}) + m_{18}(h_{18} - h_{15}) = m_L(h_{14} - h_{11})$

$m_{18} = m_4 + m_3 = 0.08782\ m_B$

Whence, substituting for m_L from eqn. (1),

$2135.4\ m_2 + 0.08782 \times 131.6\ m_B = 0.7351 \times 132.5\ m_B$

$$\therefore \frac{m_2}{m_B} = \frac{85.84}{2135.4} = \underline{0.04020}$$

Feed heater 1

$h_{11} \approx h_{12} = 270.1$ kJ/kg $h_{10} = 2392$ kJ/kg (given)

$h_9 \approx h_8 = (h_f$ at 5 kN/m²$) = 137.8$ kJ/kg

$m_1(h_{10} - h_{12}) + m_{15}(h_{15} - h_{12}) = m_L(h_{11} - h_9)$

$m_{15} = m_{18} + m_2 = (0.08782 + 0.04020)m_B = 0.12802\ m_B$

Whence, substituting for m_L from eqn. (1),

$2121.9\ m_1 + 0.12802 \times 132.5\ m_B = 0.7351 \times 132.3\ m_B$

$$\therefore \frac{m_1}{m_B} = \frac{80.29}{2121.9} = \underline{0.03784}$$

Drain flow to condenser

$m_D = m_{15} + m_1 = (0.12802 + 0.03784)m_B$

$$\therefore m_D/m_B = \underline{0.1659}$$

Steam flow to condenser

$m_C = m_L - m_D = (0.7351 - 0.1659)m_B$

$$\therefore m_C/m_B = \underline{0.5692}$$

(6) Thermal efficiency of the cycle

$$Q_{in} = m_B(h_1 - h_{29}) = (2798.3 - 966.8)m_B = 1831.5\ m_B$$

$$Q_{out} = m_C(h_7 - h_8) + m_D(h_{12} - h_8)$$

$$= [0.5692(2236.8 - 137.8) + 0.1659(270.1 - 137.8)]m_B$$

$$= (1194.8 + 21.9)m_B = 1216.7\ m_B$$

$$W_{net} = Q_{in} - Q_{out} = 614.8\ m_B$$

$$\therefore \eta_{CY} = \frac{614.8}{1831.5} \times 100 = \underline{33.6}\ \%$$

(7) Energy balance for the boiler

Primary water at inlet to boiler:

$p = 14.2\ MN/m^2$, $t = 308\ °C$, $h_{in} = 1384.3\ kJ/kg$

Primary water at outlet from boiler:

$p = 14.2\ MN/m^2$, $t = 285\ °C$, $h_{out} = 1259.2\ kJ/kg$

$$m_A(h_{in} - h_{out}) = m_B(h_1 - h_{29})$$

Here taking h_{29} as the specific enthalpy at $p = 4.4\ MN/m^2$, $t = 225\ °C$:

$$h_{29} = 967.5\ kJ/kg$$

$$\therefore \frac{m_A}{m_B} = \frac{2798.3 - 967.5}{1384.3 - 1259.2} = \underline{14.63}$$

(8) Available energy

(a) HP fluid through the reactor

Reactor outlet: $p = 14.2\ MN/m^2$, $t = 308\ °C$

$h_{out} = 1384.3\ kJ/kg$ $s_{out} = 3.3093\ kJ/kg\ K$

Reactor inlet: $p = 14.2\ MN/m^2$, $t = 285\ °C$

$h_{in} = 1259.2\ kJ/kg$ $s_{in} = 3.0896\ kJ/kg\ K$

Environment temperature = 32.9 °C $\therefore T_0 = 306.05\ K$

Per unit mass of LP fluid (steam)

Mass of HP fluid = $m_A/m_B = 14.63$

Specific available energy = $b_{out} - b_{in}$

$$= (h_{out} - h_{in}) - T_0(s_{out} - s_{in})$$

∴ <u>HP available energy</u> $= 14.63(125.1 - 306.05 \times 0.2197)$

$= 847.1$ kJ

<u>Percentage of reactor thermal output</u> $= \dfrac{847.1}{Q_{in} \text{ to steam}} \times 100$

$= \dfrac{847.1}{1831.5} \times 100 = \underline{46.3}$ %

(b) <u>LP fluid through the boiler</u>

$h_1 = 2798.3$ kJ/kg $\quad s_1 = 6.0286$ kJ/kg K

$h_{29} = 967.5$ kJ/kg $\quad s_{29} = 2.5605$ kJ/kg K

(here taking state 29 as $p = 4.4$ MN/m², $t = 225$ °C)

<u>LP available energy</u> $= (h_1 - h_{29}) - T_0(s_1 - s_{29})$

$= 1830.8 - 306.05 \times 3.4681$

$= 769.4$ kJ/kg

<u>Percentage of reactor thermal output</u> $= \dfrac{769.4}{1831.5} \times 100 = \underline{42.0}$ %

(9) <u>Minimum temperature approach (pinch point)</u>

From (7) above:

$m_A/m_B = 14.63$

Using the notation of Fig. P.8.6(b):

$h_P = 1384.3$ kJ/kg

$(h_Z - h_Y) = (h_{fg}$ at 4.4 MN/m²)

$= 1682.9$ kJ/kg

Energy balance for P to Q and Y to Z:

$m_A(h_P - h_Q) = m_B(h_Z - h_Y)$

∴ $(h_P - h_Q) = \dfrac{1682.9}{14.63} = 115.0$ kJ/kg

∴ $h_Q = 1384.3 - 115.0 = 1269.3$ kJ/kg

At $p = 14.2$ MN/m², $h = 1269.3$ kJ/kg: $\quad t_Q = 287.0$ °C

Saturation temperature at 4.4 MN/m²: $\quad t_Y = 256.0$ °C

∴ <u>Minimum temperature approach</u> $= \underline{31.0}$ K

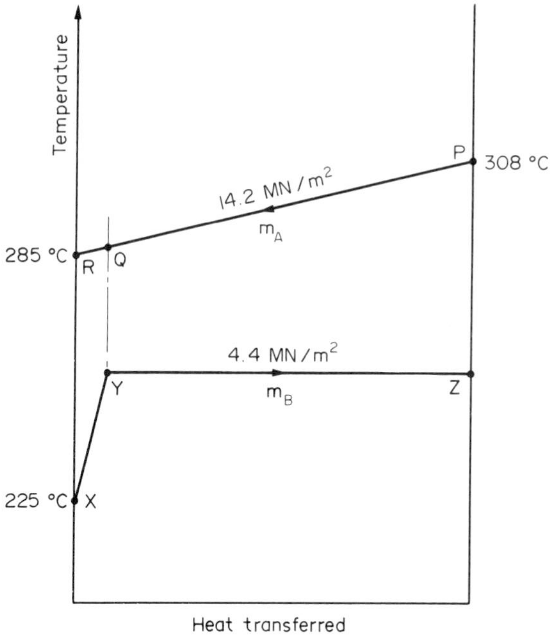

Fig. P.8.6(b)

8.7 Figure 8.13 shows a circuit diagram for a closed-circuit gas-turbine plant to be used in conjunction with a projected nuclear power plant incorporating a high-temperature, gas-cooled, fast reactor.

Sketch the temperature-entropy diagram for the corresponding ideal cycle in which a perfect gas flows round the circuit, there are no stray heat losses, no mechanical losses and no pressure drops in the heat exchangers and ducting, the expansion in the turbine is reversible and adiabatic, the compression in the compressor is reversible and isothermal, and heat exchanger X has zero terminal temperature differences, so that $T_1 = T_2$ and $T_3 = T_5$. Derive an expression for the thermal efficiency of this ideal cycle in terms of T_1, T_4, the pressure ratio of compression, r_p, and the ratio of specific heat capacities, γ.

It is estimated that, when the temperatures at points 1 and 4 are respectively 25 °C and 660 °C and $r_p = 3$, the actual thermal efficiency of a plant of this kind using helium as the circulating gas will be 35 %. Calculate the ratio of this estimated efficiency to the ideal cycle efficiency for these conditions when helium is the circulating gas.

Solution

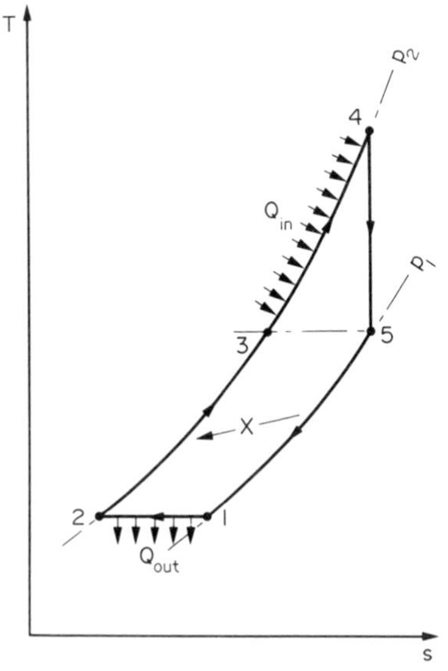

Fig. P.8.7

$$T_5 = T_4/\rho_p, \quad \text{where } \rho_p \equiv (p_2/p_1)^{(\gamma-1)/\gamma}$$

$$Q_{in} = c_p(T_4 - T_3) \quad \text{But } T_3 = T_5$$

$$\therefore\ Q_{in} = c_p\,T_4\left(1 - \frac{1}{\rho_p}\right) \tag{a}$$

For isothermal compression from 1 to 2:

T and h are both constant

$$\therefore\ T\,ds = -\,v\,dp$$

$$\text{But } \frac{v}{T} = \frac{R}{p} \text{ and } R = c_p - c_v = c_p\left(1 - \frac{1}{\gamma}\right)$$

$$\therefore\ (s_1 - s_2) = \frac{\gamma-1}{\gamma}\,c_p\,\ln r_p$$

$$Q_{out} = T_1(s_1 - s_2) = \frac{\gamma-1}{\gamma}\,c_p\,T_1\,\ln r_p \tag{b}$$

Hence, from eqns. (a) and (b):

$$\eta_{CY} = \left(1 - \frac{Q_{out}}{Q_{in}}\right) = \underline{\underline{\left[1 - \frac{T_1}{T_4}\,\frac{(\gamma-1)/\gamma}{(\rho_p-1)/\rho_p}\,\ln\,r_p\right]}}$$

For the conditions cited:

$T_1 = 25 + 273.15 = 298.15$ K $\quad T_4 = 660 + 273.15 = 933.15$ K

For helium, $\gamma = 1.67$

When $r_p = 3$, $\quad \rho_p = 3^{(0.67/1.67)} = 1.554$

$$\therefore \eta_{CY} = \left(1 - \frac{298.15}{933.15}\ \frac{0.67/1.67}{0.554/1.554} \times 1.0986\right) = 0.605$$

$\therefore$ (Actual η_{CY}/Ideal η_{CY}) = 0.35/0.605 = 0.58

8.8 Figure 8.14 shows the circuit diagram for the HTGR plant discussed in Section 8.10.1, in which the reject heat from the gas-turbine cycle is utilised to provide a heat supply for a district-heating scheme. Helium, for which $\gamma = 1.67$, circulates round the closed gas-turbine cycle, while water under pressure circulates round the closed calorifier circuit. In addition to the temperatures shown on the diagram, the following data are given:

Turbine : Pressure ratio = 2.4; isentropic efficiency = 90 %.

Compressor: Pressure ratio = 2.6; isentropic efficiency = 90 %.

The effectiveness of both the regenerative heat exchanger X and the calorifier heat exchanger is 0.87, where the effectiveness is defined as the ratio of the temperature rise of the cooler fluid to the difference between the entry temperatures of the two fluids.

Assuming that the specific heat capacities of both the helium in the gas-turbine circuit and the water in the calorifier circuit are constant, calculate the following quantities: (1) Q_P/Q_{in}, (2) the work efficiency, W_{net}/Q_{in}, (3) Q_C/Q_{in}, (4) the total efficiency, $(W_{net} + Q_C)/Q_{in}$, (5) Q_0/Q_{in}, (6) the effectiveness for which the precooler heat exchanger must be designed.

Solution

Isentropic temp. ratio, $\rho_p \equiv r_p^{(\gamma-1)/\gamma}$

Compressor

$\rho_p = 2.6^{(0.67/1.67)} = 1.4672$

$T_1 = 293.15$ K

$T_{2s} = 293.15 \times 1.4672 = 430.11$ K

$(T_{2s} - T_1) = 136.96$ K

$(T_2 - T_1) = 136.96/0.90 = 152.2$ K

$t_2 = 172.2$ °C

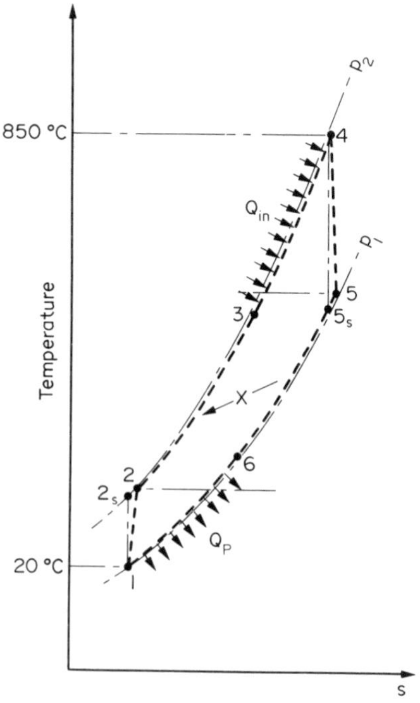

Fig. P.8.8(a)

Turbine

$$\rho_p = 2.4^{(0.67/1.67)} = 1.4208$$

$$T_4 = 1123.15\ K$$

$$T5_s = 1123.15/1.4208 = 790.51\ K$$

$$(T_4 - T5_s) = 332.64\ K$$

$$(T_4 - T_5) = 332.64 \times 0.90 = 299.4\ K$$

$$t_5 = 550.6\ °C$$

Regenerative heat exchanger X

$$(t_5 - t_2) = 550.6 - 172.2 = 378.4\ K$$

$$\text{Effectiveness} = \frac{t_3 - t_2}{t_5 - t_2} = 0.87$$

$$\therefore\ (t_3 - t_2) = 0.87 \times 378.4 = 329.2\ \text{K}$$

$$\therefore\ t_3 = 172.2 + 329.2 = 501.4\ ^\circ\text{C}$$

Treating helium as a perfect gas, so that c_p is constant,

$$(t_5 - t_6) = (t_3 - t_2) = 329.2\ \text{K}$$

$$\therefore\ t_6 = 550.6 - 329.2 = 221.4\ ^\circ\text{C}$$

(1) Calculation of Q_P/Q_{in}

$$\frac{Q_P}{Q_{in}} = \frac{t_6 - t_1}{t_4 - t_3} = \frac{221.4 - 20}{850 - 501.4} = \frac{201.4}{348.6} = \underline{0.5777}$$

(2) Work efficiency, W_{net}/Q_{in}

$$\frac{W_{net}}{Q_{in}} = \left(1 - \frac{Q_P}{Q_{in}}\right) = (1 - 0.5777) = \underline{0.4223}$$

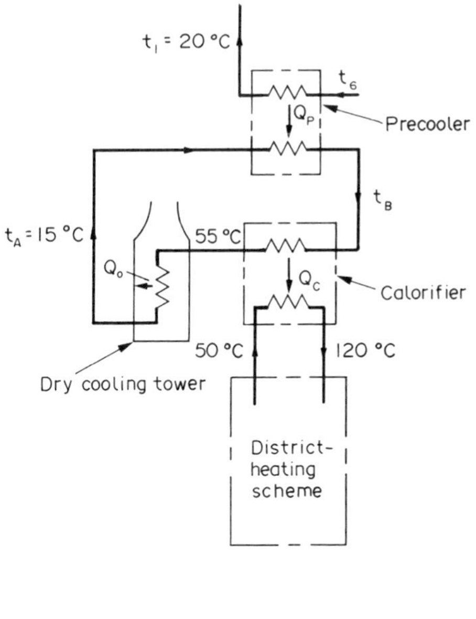

Fig. P.8.8(b)

(3) Calculation of Q_C/Q_{in}

Calorifier heat exchanger

Effectiveness = 0.87

$$\therefore\ (t_B - 50) = (120 - 50)/0.87$$

$$t_B = 130.5\ ^\circ C$$

$$Q_C/Q_P = \frac{130.5 - 55}{130.5 - 15} = 0.6537$$

$$\frac{Q_C}{Q_{in}} = \frac{Q_C}{Q_P}\frac{Q_P}{Q_{in}} = 0.6537 \times 0.5777 = \underline{0.3776}$$

(4) Total efficiency, $(W_{net} + Q_C)/Q_{in}$

$$(W_{net} + Q_C)/Q_{in} = 0.4223 + 0.3776 = \underline{0.800}$$

(5) Calculation of Q_0/Q_{in}

$$\frac{Q_0}{Q_P} = \frac{55 - 15}{t_B - 15} = \frac{40}{115.5} = 0.3463$$

$$\frac{Q_0}{Q_{in}} = \frac{Q_0}{Q_P}\frac{Q_P}{Q_{in}} = 0.3463 \times 0.5777 = \underline{0.200}$$

Note that this is also equal to (1 - Total efficiency).

(6) Precooler heat exchanger

$$\text{Effectiveness} = \frac{t_B - 15}{t_6 - 15} = \frac{130.5 - 15}{221.4 - 15} = \underline{0.560}$$

CHAPTER 9

Combined and Binary Plant

9.1 For the ideal super-regenerative steam cycle illustrated in Figs 9.1 and 9.2, show that

$$\frac{M_B}{M_G} = \frac{\Delta s_3}{\Delta s_2} \text{ and } \frac{M_A}{M_G} = \frac{\Delta s_3}{\Delta s_1}.$$

In an "ideal" super-regenerative steam cycle, evaporation takes place at 10 MN/m², the maximum cycle temperature is 550 °C, the exit pressure from the reheated turbine is 1.0 MN/m² and the condenser pressure is 4.0 kN/m². Determine:

(a) M_B/M_G and M_A/M_G,

(b) the thermal efficiency of the cycle,

(c) the Carnot efficiency for the same extreme temperature limits.

Noting that the difference between the Carnot efficiency and the cycle efficiency arises from two causes, namely (1) the fact that the heat to the cycle is not all added at the top temperature, and (2) the irreversibility of the process occurring in heat exchanger X of Figs. 9.1 and 9.2, evaluate:

(d) the entropy creation due to irreversibility (Section A.6 of Appendix A) in each of these processes, expressed per unit mass of steam passing through X,

(e) the fraction of the total loss in cycle efficiency below η_{Carnot} that is attributable to each of these causes.

Solution

Using the notation of Fig. P.9.1, which is identical to Fig. 9.2 in Ref. 1:

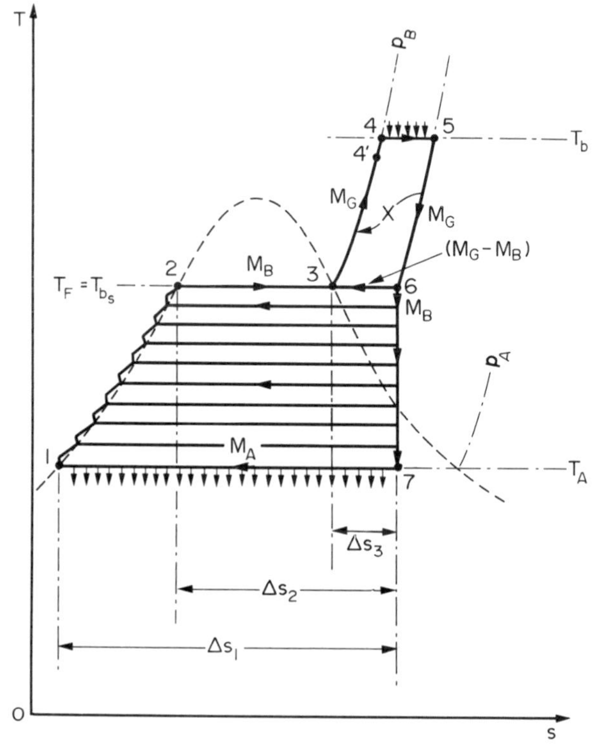

Fig. P.9.1 (As Fig. 9.2)

<u>For the intercooling process</u> in the saturated steam drum of Fig. 9.1 in Ref. 1:

$$(M_G - M_B)(h_6 - h_3) = M_B(h_3 - h_2) \quad (1)$$

$$(M_G - M_B)(s_6 - s_3)T_3 = M_B(s_3 - s_2)T_3$$

$$(M_G - M_B)\Delta s_3 = M_B(\Delta s_2 - \Delta s_3)$$

$$\therefore \frac{M_B}{M_G} = \frac{\Delta s_3}{\Delta s_2} \quad (2)$$

From equation (1) above:

$$M_G(h_6 - h_3) = M_B(h_6 - h_2)$$

Thus, the complete cycle is equivalent to a "gas-turbine" cycle 34563 rejecting heat $M_G(h_6 - h_3)$ to a completely reversible regenerative steam cycle 12671 receiving an equal quantity of heat $M_B(h_6 - h_2)$. Not only is the hypothetical cycle 12671 completely reversible, but all heat is received at temperature T_F and all heat is rejected at temperature T_A, so that it has Carnot cycle efficiency. Hence, for the hypothetical cycle 12671,

$$(1 - \eta_{CY}) = \frac{T_A}{T_F} \qquad (3)$$

But $$(1 - \eta_{CY}) = \frac{Q_{out}}{Q_{in}} = \frac{M_A(h_7 - h_1)}{M_G(h_6 - h_3)} = \frac{M_A T_A \Delta s_1}{M_G T_F \Delta s_3} \qquad (4)$$

From eqns. (3) and (4), $$\underline{\underline{\frac{M_A}{M_G} = \frac{\Delta s_3}{\Delta s_1}}} \qquad (5)$$

The states of the fluid at the points depicted in Fig. P.9.1 are as follows:

$p_B = 10\ \text{MN/m}^2, \quad t_4 = 550\ °\text{C}$

$h_4 = 3500\ \text{kJ/kg} \quad s_4 = 6.756\ \text{kJ/kg K}$

$h_3 = 2727.7\ \text{kJ/kg} \quad s_3 = 5.620\ \text{kJ/kg K}$

$h_2 = 1408.0\ \text{kJ/kg} \quad s_2 = 3.361\ \text{kJ/kg K}$

$p_5 = p_6 = 1.0\ \text{MN/m}^2, \quad t_5 = 550\ °\text{C}$

$h_5 = 3587\ \text{kJ/kg} \quad s_5 = 7.899\ \text{kJ/kg K}$

$t_6 = t_3 =$ Saturation temperature at $10\ \text{MN/m}^2 = 311.0\ °\text{C}$

$h_6 = 3075.8\ \text{kJ/kg} \quad s_7 = s_6 = 7.165\ \text{kJ/kg K}$

$p_1 = p_7 = p_A = 4.0\ \text{kN/m}^2$

$h_1 = 121.4\ \text{kJ/kg} \quad s_1 = 0.422\ \text{kJ/kg K}$

$\Delta s_1 \equiv (s_7 - s_1) = 6.743\ \text{kJ/kg K}$

$\Delta s_2 \equiv (s_7 - s_2) = 3.804\ \text{kJ/kg K}$

$\Delta s_3 \equiv (s_7 - s_3) = 1.545\ \text{kJ/kg K}$

(a) Calculation of flow ratios

$$\frac{M_B}{M_G} = \frac{\Delta s_3}{\Delta s_2} = \frac{1.545}{3.804} = \underline{0.406}$$

$$\frac{M_A}{M_G} = \frac{\Delta s_3}{\Delta s_1} = \frac{1.545}{6.743} = \underline{0.229}$$

(b) Calculation of thermal efficiency

Note that, although the specified cycle is described as "ideal", it fails to come up fully to the requirements of the theoretical ideal cycle because of the irreversibilities mentioned in the question and studied shortly in (d) and (e).

Because the mean specific heat capacity of superheated steam is greater at 10 MN/m² than at 1.0 MN/m², the steam is heated to something lower than 550 °C on the high-pressure side of heat exchanger X, namely to state point 4' instead of state point 4. To determine the temperature at 4', the energy conservation equation is applied to heat exchanger X. Thus:

$$h_{4'} - h_3 = h_5 - h_6$$

$$\therefore \quad h_{4'} = 2727.7 + (3587 - 3075.8) = 3239 \text{ kJ/kg}$$

$$\text{At 10 MN/m}^2, \; t_{4'} = 448.2 \text{ °C}$$

$$s_{4'} = 6.417 \text{ kJ/kg K}$$

From 4' to 5:

$$Q_{in} = [(h_4 - h_{4'}) + T_4(s_5 - s_4)]M_G$$

$$= (261 + 823.15 \times 1.143)M_G = 1202\ M_G \text{ kJ}$$

From 7 to 1:

$$Q_{out} = [T_A(s_7 - s_1)]M_A$$

$$= (302.15 \times 6.743)M_A = 2037.4\ M_A$$

$$\eta_{CY} = \left(1 - \frac{Q_{out}}{Q_{in}}\right)$$

$$= \left(1 - \frac{2037.4}{1202}\frac{M_A}{M_G}\right)$$

$$\therefore \quad \eta_{CY} = \left(1 - \frac{2037.4}{1202} \times 0.229\right) \times 100 = \underline{61.2}\ \%$$

(c) Calculation of Carnot efficiency

$$\eta_{CARNOT} = \left(1 - \frac{T_A}{T_b}\right)$$

$$= \left(1 - \frac{302.15}{823.15}\right) \times 100 = \underline{63.3}\ \%$$

(d) Calculation of entropy creation quantities

In the heating process 4' to 4, heat is transferred to the steam from a source at 550 °C (823.15 K).

Entropy creation, $\Delta S_C \equiv \Delta S - \Delta S_Q$, where ΔS_Q is the thermal entropy flux brought into the fluid from the heat source at 550 °C. [See Appendix A, eqn. (A.9) of Ref. 1.]

$$\therefore \Delta S_C = (s_4 - s_{4'}) - \frac{h_4 - h_{4'}}{T_4}$$

$$= 0.339 - \frac{261}{823.15} = \underline{0.022} \text{ kJ/kg K}$$

In the heat exchanger X, when stray heat loss is assumed to be negligible, so that $\Delta S_Q = 0$,

$$\Delta S_C = (s_{4'} - s_3) - (s_5 - s_6)$$

$$= 0.797 - 0.734 = \underline{0.063} \text{ kJ/kg K}$$

(e) Calculation of lost work due to irreversibility

The loss in cycle efficiency due to irreversibility is proportional to the lost work due to irreversibility, which itself is proportional to the entropy creation due to irreversibility, since

$$\text{Lost work} = T_0 \, \Delta S_C,$$

where T_0 is the temperature of the conceptual environment.

$$\text{Total entropy creation} = 0.022 + 0.063 = 0.085 \text{ kJ/kg K}$$

$$\frac{\text{Loss due to heating process}}{\text{Total loss}} = \frac{0.022}{0.085} = \underline{0.26}$$

$$\frac{\text{Loss due to exchanger X}}{\text{Total loss}} = \frac{0.063}{0.085} = \underline{0.74}$$

9.2 At the design load of a Field super-regenerative steam cycle such as that illustrated in Figs. 9.3 and 9.4, the steam flow rate M_B to the condensing turbine is such that the quantity of feed water injected into the spray desuperheater is just sufficient to ensure that the condition of the steam at the compressor outlet is dry saturated. It may be assumed that there is complete mixing of the injected water and the steam before entry to the compressor. The pressures and temperatures at the points specified are:

Position	Pressure (MN/m²)	Temperature (°C)
2	0.7	100
4	7.0	(dry saturated)
6	7.0	550
8	2.5	550
10	0.7	300
11	0.004	(wet)

The enthalpy rise of the feed water is the same in both direct-contact heaters, and β (as defined in Section 7.9) may be taken as equal to 2270 kJ/kg in each. The isentropic efficiency of each turbine, and of the compressor, is 85 %. Pressure drops in the piping may be neglected.

Determine the ratios M_B/M_G and M_B/M_A. Calculate the thermal efficiencies η_G, η_S and η_F as defined in Section 9.4, and check that they satisfy eqn. (9.6).

Solution

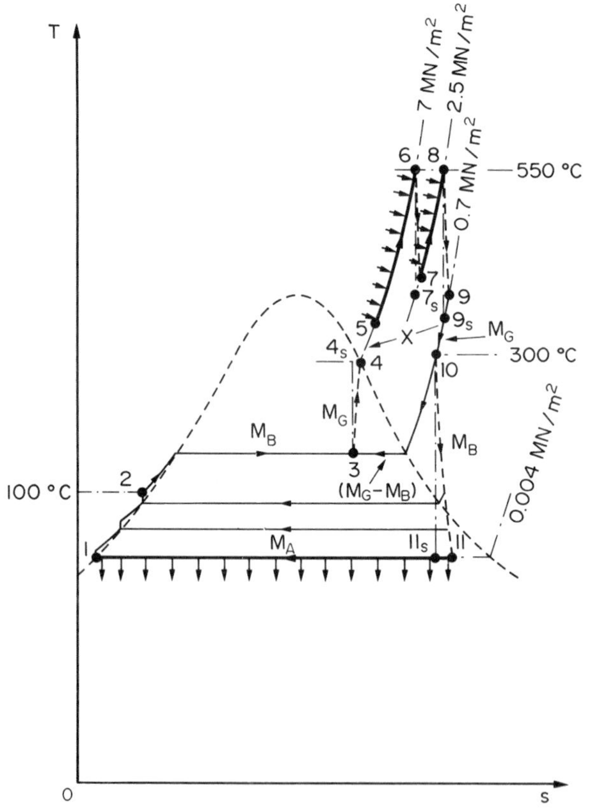

Fig. P.9.2 (Similar to Fig. 9.4)

Using the enthalpy-entropy diagram for steam provided with Ref. 2, and the notation of Fig. P.9.2:

HP turbine

$$h_6 = 3530 \text{ kJ/kg}$$

$$h_{7_s} = 3195 \text{ kJ/kg}$$
$$\Delta h_s = 335 \text{ kJ/kg}$$
$$\Delta h = 335 \times 0.85$$
$$= 284.8 \text{ kJ/kg}$$
$$h_7 = 3245.2 \text{ kJ/kg}$$

LP turbine

$$h_8 = 3573 \text{ kJ/kg}$$
$$h_{9_s} = 3158 \text{ kJ/kg}$$
$$\Delta h_s = 415 \text{ kJ/kg}$$
$$\Delta h = 415 \times 0.85$$
$$= 352.8 \text{ kJ/kg}$$
$$h_9 = 3220.2 \text{ kJ/kg}$$

Compressor

At 7 MN/m², $s_g = 5.816$ kJ/kg K $\quad s_{fg} = 2.694$ kJ/kg K

$h_g = 2773.5$ kJ/kg $\quad h_{fg} = 1506.0$ kJ/kg

At 0.7 MN/m², $s_g = 6.705$ kJ/kg K $\quad s_{fg} = 4.713$ kJ/kg K

$h_g = 2762.0$ kJ/kg $\quad h_{fg} = 2064.9$ kJ/kg

To determine state point 3 at compressor inlet

At inlet, $(1 - x_3) = \dfrac{6.705 - s_3}{4.713}$

$$h_3 = 2762.0 - (1 - x_3)2064.9$$
$$= 2762.0 - \left(\frac{6.705 - s_3}{4.713}\right)2064.9$$
$$\therefore h_3 = 438.1\, s_3 - 175.7 \qquad (1)$$
$$h_4 = 2773.5$$
$$\Delta h \equiv (h_4 - h_3) = 2949.2 - 438.1\, s_3$$

At exit, $(1 - x_{4_s}) = \dfrac{5.816 - s_3}{2.694}$, since $s_{4_s} = s_3$

$$h_{4_s} = 2773.5 - (1 - x_{4_s})1506.0$$
$$= 2773.5 - \left(\frac{5.816 - s_3}{2.694}\right)1506.0$$

$$\therefore h_{4_s} = 559.0\ s_3 - 477.8$$

$$\Delta h_s \equiv (h_{4_s} - h_3) = 120.9\ s_3 - 302.1$$

Compressor efficiency

$$\eta_C \equiv \frac{\Delta h_s}{\Delta h} = \frac{120.9\ s_3 - 302.1}{2949.2 - 438.1\ s_3} = 0.85$$

$$\text{Whence } s_3 = \frac{2808.9}{493.3} = 5.694 \text{ kJ/kg K} \qquad (2)$$

From (1) and (2):

$$h_3 = 438.1 \times 5.694 - 175.7 = 2318.8 \text{ kJ/kg}$$

Calculation of M_B/M_G

At 0.7 MN/m² and 300 °C, $h_{10} = 3060$ kJ/kg, $s_{10} = 7.300$ kJ/kg K

$h_2 \approx h_f$ at 100 °C = 419.1 kJ/kg

In the spray desuperheater:

$$(M_G - M_B)h_{10} + M_B h_2 = M_G h_3$$

$$\therefore \frac{M_B}{M_G} = \frac{h_{10} - h_3}{h_{10} - h_2} = \frac{3060 - 2318.8}{3060 - 419.1} = \underline{0.281}$$

Calculation of M_B/M_A

At 0.004 MN/m², $h_g = 2554.5$ kJ/kg, $s_g = 8.475$ kJ/kg K

$h_1 = h_f = 121.4$ kJ/kg, $s_f = 0.422$ kJ/kg K

In the feed heaters:

Total enthalpy rise $R \approx (h_2 - h_1) = 419.1 - 121.4 = 297.7$ kJ/kg

From equations (7.7) and (7.8) in Chapter 7 of Ref. 1, for a system of n heaters, with equal enthalpy rises and equal values of β in all heaters:

$$\frac{\text{Exit flow rate}}{\text{Inlet flow rate}} = \left(1 + \frac{R}{n\beta}\right)^n$$

Hence, for 2 heaters, and with β = 2270 kJ/kg (given):

$$\frac{M_B}{M_A} = \left(1 + \frac{297.7}{2 \times 2270}\right)^2 = \underline{1.135}$$

<u>To determine state point 5 at exit from regenerative heat exchanger X</u>

In heat exchanger X:

$$(h_5 - h_4) = (h_9 - h_{10})$$

$$h_5 - 2773.5 = 3220.2 - 3060$$

$$\therefore h_5 = 2933.7 \text{ kJ/kg}$$

<u>Hypothetical "gas-turbine" cycle 3693:</u>

$$(Q_{in})_G = (h_6 - h_5) + (h_8 - h_7)$$

$$= (3530 - 2933.7) + (3573 - 3245.2)$$

$$= 924.1 \text{ kJ/kg}$$

$$(Q_{out})_G = (h_{10} - h_3)$$

$$= 3060 - 2318.8 = 741.2 \text{ kJ/kg}$$

$$\therefore \eta_G = \left(1 - \frac{741.2}{924.1}\right) \times 100 = \underline{19.8} \ \%$$

<u>Hypothetical regenerative condensing steam cycle, 1-2-10-11-1</u>

<u>Steam turbine 10-11</u>

$$(1 - x_{11_s}) = \frac{s_g - s_{10}}{s_g - s_f} = \frac{8.475 - 7.300}{8.475 - 0.422} = 0.1459$$

$$h_{11_s} = h_g - (1 - x_{11_s})h_{fg}$$

$$= 2554.5 - 0.1459 \times 2433.1 = 2199.5 \text{ kJ/kg}$$

$$\Delta h_s = (h_{10} - h_{11_s}) = 3060 - 2199.5 = 860.5 \text{ kJ/kg}$$

$$\Delta h = 860.5 \times 0.85 = 731.4 \text{ kJ/kg}$$

$$h_{11} = 3060 - 731.4 = 2328.6 \text{ kJ/kg}$$

Hypothetical steam cycle

$$(Q_{in})_S = M_B(h_{10} - h_2)$$

$$(Q_{out})_S = M_A(h_{11} - h_1)$$

$$\therefore\ \eta_S = \left[1 - \frac{M_A}{M_B}\frac{(h_{11} - h_1)}{(h_{10} - h_2)}\right]$$

$$= \left[1 - \frac{1}{1.135}\frac{2328.6 - 121.4}{3060 - 419.1}\right] \times 100 = \underline{26.4}\ \%$$

Field cycle

$$(Q_{in})_F = M_G(Q_{in})_G$$

$$(Q_{out})_F = M_A(h_{11} - h_1)$$

$$\therefore\ \eta_F = \left[1 - \frac{M_A}{M_G}\frac{(h_{11} - h_1)}{(Q_{in})_G}\right]$$

$$= \left[1 - \frac{M_B/M_G}{M_B/M_A}\frac{(h_{11} - h_1)}{(Q_{in})_G}\right]$$

$$= \left[1 - \frac{0.281}{1.135}\frac{2207.2}{924.1}\right] \times 100 = \underline{40.9}\ \%$$

Check against eqn. (9.6) of Ref. 1:

Eqn. (9.6) is $(1 - \eta_F) = (1 - \eta_G)(1 - \eta_S)$

$$= (1 - 0.198)(1 - 0.264) = 0.590$$

$$\therefore\ \eta_F = (1 - 0.590) \times 100 = \underline{41.0}\ \%$$

Subject to rounding error, this is the same as the previous value.

9.3 Derive eqn. (9.15) for the *exhaust-heated* combined gas-steam plant described in Section 9.6, the relevant calorific value being the combined heating value of the fuel burnt in the two combustion chambers.

Solution

Fig. P.9.3 depicts a simplified flow diagram for the exhaust-heated combined gas-steam plant (with a flue-gases after-cooler) described in Section 9.6.

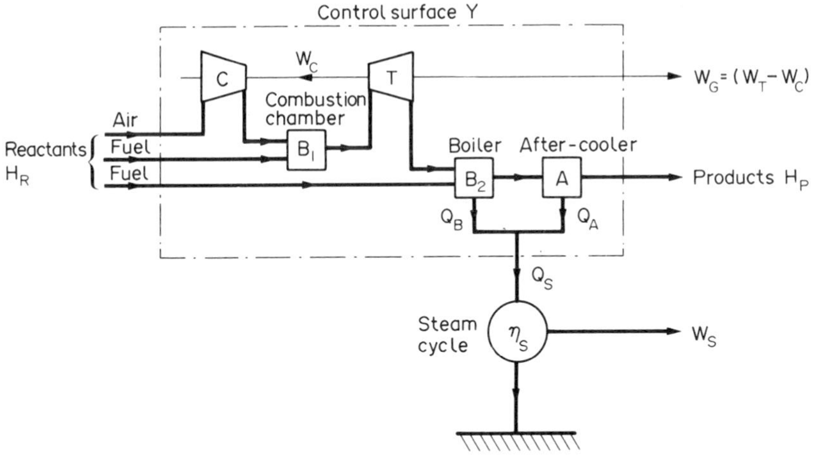

Fig. P.9.3

The steady-flow energy equation for control surface Y is:

$$Q_S + W_G = (H_R - H_P) = \eta_B \, CV$$

where CV is the combined heating value of the fuel burnt in the two combustion chambers, B_1 and B_2.

$$\text{Writing } x \equiv \frac{W_G}{CV} \qquad (1)$$

$$Q_S = (\eta_B - x)\, CV \qquad (2)$$

$$\text{while } W_S = \eta_S \, Q_S \qquad (3)$$

The overall efficiency of the complete plant, neglecting the power required for auxiliaries, is defined by

$$\eta_O \equiv \frac{W_G + W_S}{CV} \qquad (4)$$

Whence, from equations (1), (2), (3) and (4):

$$\eta_O = x + \eta_S(\eta_B - x) \qquad (5)$$

By writing $\eta_O' \equiv \eta_B \eta_S$ (6)

equation (5) becomes

$$\eta_O = \eta_O' + x(1 - \eta_S) \qquad (9.15)$$

9.4 In a mercury-steam binary vapour cycle the mercury leaves the mercury boiler dry-saturated at 900 kN/m², and is condensed in the mercury-steam condenser-boiler at 17 kN/m². Dry saturated steam leaves the condenser-boiler at 4 MN/m², and is heated to 425 °C in a superheater placed in the flue-gas passes of the mercury boiler.

The steam is supplied to a regenerative steam cycle in which the condenser pressure is 4 kN/m². The feed water from the condenser is heated to 170 °C in the three bled-steam feed-heating stages, and is then further heated to the steam saturation temperature by being passed through an economiser placed in the exit flue gas stream from the mercury boiler. For the purposes of calculation it may be assumed that three direct-contact feed heaters are used, that the enthalpy rise of the feed water is the same in each, and also that β (as defined in Section 7.9) is equal to 2215 kJ/kg in each heater. The isentropic efficiency of the mercury turbine is 75 % and of the steam turbine is 82 %.

The properties of saturated mercury are given below:

Pressure / kN/m²	Enthalpy/(kJ/kg)		Entropy/(kJ/kg K)	
	h_f	h_g	s_f	s_g
900	68.4	358.1	0.1415	0.5121
17	36.9	331.1	0.0940	0.6362

Calculate:

(a) the ratio of the mass flow rates of steam through the condenser-boiler and steam condenser respectively;

(b) the ratio of the mass flow rates of mercury and steam through the condenser-boiler;

(c) the thermal efficiency η_{CY} of the binary cycle;

(d) the overall efficiency η_O and heat rate of the plant (in Btu/kW h), given that the efficiency of the mercury boiler is 85 %;

(e) the thermal efficiencies η_M and η_S defined in Section 9.11;

(f) the value of q defined by eqn. (9.27); check that η_{CY} satisfies eqn. (9.26);

(g) the value of x defined by eqn. (9.31); check that η_O satisfies eqn. (9.29).

Note: There was an error in the original framing of (e), so that (e) and (f) have been revised and (g) has been added.

Solution

Mercury cycle

After isentropic expansion from state C:

$$(1 - x_D) = \frac{0.6362 - 0.5121}{0.6362 - 0.0940} = 0.2289$$

$$h_D = 331.1 - 0.2289 \times 294.2 = 263.8 \text{ kJ/kg}$$

$$\Delta h_s \equiv (h_C - h_D) = 358.1 - 263.8 = 94.3 \text{ kJ/kg}$$

$$\Delta h = \eta_T \Delta h_S = 0.75 \times 94.3 = 70.7 \text{ kJ/kg}$$

$$h_E = (h_C - \Delta h) = 358.1 - 70.7 = 287.4 \text{ kJ/kg}$$

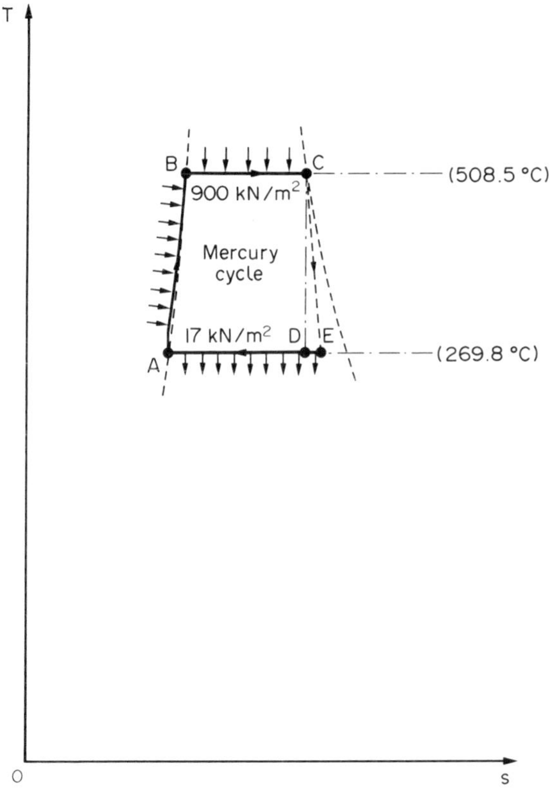

Fig. P.9.4(a)

Steam cycle

At 4 kN/m²:

h_g = 2554.5 kJ/kg s_g = 8.475 kJ/kg K

h_f = 121.4 kJ/kg s_f = 0.422 kJ/kg K

After isentropic expansion from state 5:

$$s_6 = s_5 = 6.858 \text{ kJ/kg K}$$

$$(1 - x_6) = \frac{8.475 - 6.858}{8.475 - 0.422} = 0.2008$$

$$h_6 = 2554.5 - 0.2008 \times 2433.1 = 2065.9 \text{ kJ/kg}$$

$$h_5 = 3274 \text{ kJ/kg}$$

$$\Delta h_s \equiv (h_5 - h_6) = 3274 - 2065.9 = 1208.1 \text{ kJ/kg}$$

$$\Delta h = \eta_T \Delta h_s = 0.82 \times 1208.1 = 990.6 \text{ kJ/kg}$$

$$h_7 = (h_5 - \Delta h) = 3274 - 990.6 = 2283.4 \text{ kJ/kg}$$

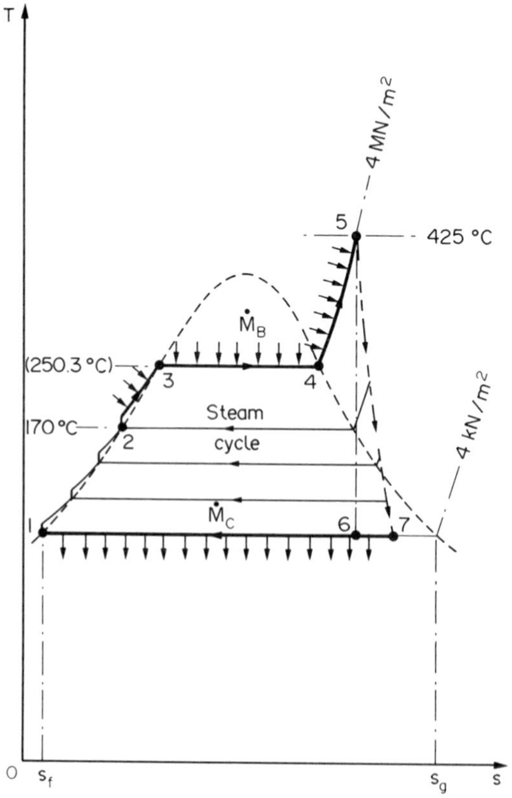

Fig. P.9.4(b)

(a) Ratio of steam flow rates, $\dot{M}_B/\dot{M}_C$

At 170 °C, $h_2 = h_f = 719.1$ kJ/kg

At 4 kN/m², $h_1 = h_f = \underline{121.4}$ kJ/kg

In the feed heaters:

Total enthalpy rise, $R = 597.7$ kJ/kg

From equations (7.7) and (7.8) in Chapter 7 of Ref. 1:

$$\frac{\dot{M}_B}{\dot{M}_C} = \left(1 + \frac{R}{n\beta}\right)^n$$

where n = 3

$\beta = 2215$ kJ/kg (given)

$$\therefore \frac{\dot{M}_B}{\dot{M}_C} = \left(1 + \frac{597.7}{3 \times 2215}\right)^3 = \underline{1.295}$$

(b) Ratio of flow rates of mercury and steam

Let the mass flow rate of mercury = $\dot{M}_M$

In the mercury-steam condenser-boiler:

From the mercury, $\dot{Q}_{out} = \dot{M}_M(h_E - h_A)$

$= (287.4 - 36.9)\dot{M}_M = 250.5\ \dot{M}_M$

To the steam, $\dot{Q}_{in} = \dot{M}_B(h_4 - h_3)$

At 4 MN/m², $h_4 = h_g = 2800.3$ kJ/kg

$h_3 = h_f = 1087.4$ kJ/kg

$\therefore \dot{Q}_{in} = (2800.3 - 1087.4)\dot{M}_B = 1712.9\ \dot{M}_B$

But $(\dot{Q}_{in})_{steam} = (\dot{Q}_{out})_{mercury}$

$$\therefore \frac{\dot{M}_M}{\dot{M}_B} = \frac{1712.9}{250.5} = \underline{6.84}$$

(c) Thermal efficiency, η_{CY}, of the binary cycle

Per kg of steam delivered by the condenser-boiler:

Steam cycle

Total $Q_{in} \approx (h_5 - h_2) = 3274 - 719.1 = 2554.9$ kJ

$$Q_{out} = \frac{(h_7 - h_1)}{\dot{M}_B/\dot{M}_C} = \frac{2283.4 - 121.4}{1.295} = \underline{1669.5} \text{ kJ}$$

$$W_{net} = (Q_{in} - Q_{out}) = 885.4 \text{ kJ}$$

Mercury cycle

Neglecting the feed pump work input:

Net work output = Δh = 70.7 kJ per kg of mercury

$$\therefore W_{net} = \frac{\dot{M}_M}{\dot{M}_B} \times 70.7 = 6.84 \times 70.7 = 483.6 \text{ kJ}$$

Binary cycle

Total W_{net} = 885.4 + 483.6 = 1369.0 kJ

Q_{in} from the fuel:

In the steam-cycle economiser, $Q_{in} \approx (h_3 - h_2)$

$= (1087.4 - 719.1)$

$= 368.3$ kJ

In the steam-cycle superheater, $Q_{in} = (h_5 - h_4)$

$= (3274 - 2800.3)$

$= 473.7$ kJ

Using the notation of Fig. 9.11 in Chapter 9 of Ref. 1:

$Q_3 \equiv$ Total Q_{in} from the fuel to the steam cycle

$= 368.3 + 473.7 = 842.0$ kJ

$$Q_1 \equiv Q_{in} \text{ to mercury cycle} \approx \frac{\dot{M}_M}{\dot{M}_B}(h_C - h_A) = 6.84(358.1 - 36.9)$$

$= 6.84 \times 321.2$

$= 2197.0$ kJ

Total Q_{in} from the fuel $= (Q_1 + Q_3) = (842.0 + 2197.0)$

$= 3039.0$ kJ

$$\therefore \; \eta_{CY} = \frac{\text{Total } W_{net}}{(Q_1 + Q_3)} = \frac{1369.0}{3039.0} \times 100 = \underline{45.1}\ \%$$

(d) Overall efficiency, η_O, and heat rate

$$\eta_O = \eta_B\, \eta_{CY} = 0.85 \times 0.451 \times 100 = \underline{38.3}\ \%$$

1 kW h ≈ 3412 Btu

$$\text{Heat rate} \equiv \frac{Q_{in}}{W_{net}} = \frac{1}{\eta_O} \text{ kJ/kJ}$$

$$\therefore \text{Heat rate} = \frac{3412}{0.383} = \underline{8910} \text{ Btu/kW h}$$

(e) Thermal efficiencies η_M and η_S

Mercury cycle

From (c) above:

$$\eta_M \equiv \frac{W_{net}}{Q_{in}} = \frac{483.6}{2197.0} \times 100 = \underline{22.0}\ \%$$

Hypothetical steam cycle

From (c) above:

$$\eta_S \equiv \frac{W_{net}}{\text{Total } Q_{in}} = \frac{885.4}{2554.9} \times 100 = \underline{34.7}\ \%$$

(f) Calculation of q and checking of η_{CY} against eqn. (9.26)

Using the notation of Fig. 9.11 in Chapter 9 of Ref. 1:

$$q \equiv \frac{Q_1}{Q_1 + Q_3} = \frac{Q_{in} \text{ to mercury cycle}}{\text{Total } Q_{in} \text{ from the fuel}}$$

$$\therefore \text{ From (c) above, } q = \frac{2197.0}{3039.0} = \underline{0.723}$$

Checking of η_{CY} against equation (9.26)

$$(1 - \eta_{CY}) = (1 - q\,\eta_M)(1 - \eta_S) \tag{9.26}$$

$$= (1 - 0.723 \times 0.220)(1 - 0.347)$$

$$= 0.841 \times 0.653 = 0.549$$

$$\therefore \eta_{CY} = 0.451 \times 100 = \underline{45.1}\ \%$$

This agrees with the value of η_{CY} calculated in (c) above

(g) <u>Calculation of x and checking of η_o against equation (9.29)</u>

$$x \equiv \frac{W_M}{CV} = q\ \eta_M\ \eta_B \qquad (9.31)$$

$$\therefore x = 0.723 \times 0.220 \times 0.85 = \underline{0.135}$$

$$\eta'_o \equiv \eta_B\ \eta_S = 0.85 \times 0.347 = 0.295$$

$$\eta_o = \eta'_o + x(1 - \eta_S) \qquad (9.29)$$

$$\therefore \eta_o = (0.295 + 0.135 \times 0.653) \times 100 = \underline{38.3}\ \%$$

This agrees with the value of η_o calculated in (d) above

9.5 Derive equation (9.25)

Solution

To show that, for the <u>ideal</u> mercury-steam binary cycle depicted in Fig. 9.10 in Chapter 9 of Ref. 1,

$$(1 - \eta_{CY}) = (1 - \eta_M)(1 - \eta_S) \qquad (9.25)$$

For the binary cycle, $(1 - \eta_{CY}) = \dfrac{Q_{41}}{Q_{AC}}$

For the mercury cycle, $(1 - \eta_M) = \dfrac{Q_{DA}}{Q_{AC}}$

For the steam cycle, $(1 - \eta_S) = \dfrac{Q_{41}}{Q_{AD}}$

But $Q_{AD} = Q_{DA}$

$$\therefore (1 - \eta_M)(1 - \eta_S) = \frac{Q_{41}}{Q_{AC}} = (1 - \eta_{CY})$$

CHAPTER 10

Advanced Refrigerating and Gas-Liquefaction Plant

10.1 In an ammonia absorption refrigerating plant (Fig. 10.1), the absolute pressures in the condenser and evaporator are respectively 1.17 MN/m^2 and 0.24 MN/m^2. Stray heat transfers with the environment and pressure drops in all items of the plant other than across the throttle valves may be neglected. The plant is not fitted with the heat exchanger X shown in the figure.

(a) A saturated solution of ammonia and H_2O liquid leaves the absorber at E at 32 °C. A saturated mixture of ammonia and H_2O vapour leaves the generator at A, and a saturated solution of ammonia and H_2O liquid leaves at H, both streams being at 105 °C. Calculate, per kg of mixture leaving the generator at A, (i) the quantity of aqua-ammonia solution entering the generator, (ii) the quantity of heat Q_3 supplied to the generator, (iii) the fraction of Q_3 that is used to raise the temperature of the weak solution returned to the absorber.

(b) An aqua-ammonia solution leaves the condenser at 32 °C. Calculate the quantity of heat Q_2 transferred to the cooling water passing through the condenser.

(c) The mixture leaves the evaporator at -7 °C. Calculate the liquid fraction at D, and the quantity of heat Q_1 supplied to the evaporator from the cold chamber.

(d) Calculate, per kg of mixture entering the absorber, the quantity of heat Q_2' transferred to the cooling water passing through the absorber.

(e) Calculate the coefficient of performance of the plant, neglecting the work input to the pump.

(f) Calculate the coefficient of performance of the corresponding ideal plant in which T_1, T_2 and T_3 are the absolute temperatures corresponding respectively to the temperatures of the aqua ammonia leaving the evaporator, condenser (and absorber) and generator. Calculate also the coefficient of performance of a reversed-Carnot refrigerator operating between the same values of T_1 and T_2.

The equilibrium properties of aqua ammonia are given in the following table. The enthalpy of a liquid mixture of given concentration and temperature may be assumed to be independent of pressure.

Pressure (MN/m^2):	0.24		0.24	1.17	
Temperature (°C):	-7		32	105	
	Liquid	Vapour	Liquid	Liquid	Vapour
Concentration x (kg NH_3/kg mixture)	0.755	0.999	0.402	0.318	0.925
Specific enthalpy h (kJ/kg)	-200	1270	-110	280	1570

The specific enthalpy of an aqua-ammonia solution of concentration 0.925 at 32 °C is 100 kJ/kg.

Solution

(a) Vapour generator

(i) Let m = quantity of mixture entering at G

Then a mass balance for NH_3 gives:

$$m \times 0.402 = 1 \times 0.925 + (m - 1) \times 0.318$$

$$\therefore m = \frac{0.607}{0.084} = \underline{7.23} \text{ kg}$$

(ii) Quantity of mixture leaving at H = 7.23 - 1 = 6.23 kg

The energy balance for the vapour generator gives:

$$Q_3 = (1 \times 1570 + 6.23 \times 280) - 7.23 \times (-110)$$

$$\therefore Q_3 = 1570 + 1744 + 795 = \underline{4109} \text{ kJ}$$

(iii) Heat quantity to the weak solution

$$= 6.23[280 - (-110)] = 2430 \text{ kJ}$$

$$\text{Fraction of } Q_3 = \frac{2430}{4109} = \underline{0.591}$$

(b) Condenser

$$Q_2 = (h_A - h_B) = 1570 - 100 = \underline{1470}\ kJ$$

(c) Evaporator

Let y = liquid fraction leaving the evaporator.

Then, per kg of mixture, a mass balance for NH_3 gives:

$$0.755\,y + 0.999(1 - y) = 0.925 \times 1$$

$$\therefore\ y = \frac{0.074}{0.244} = \underline{0.303}$$

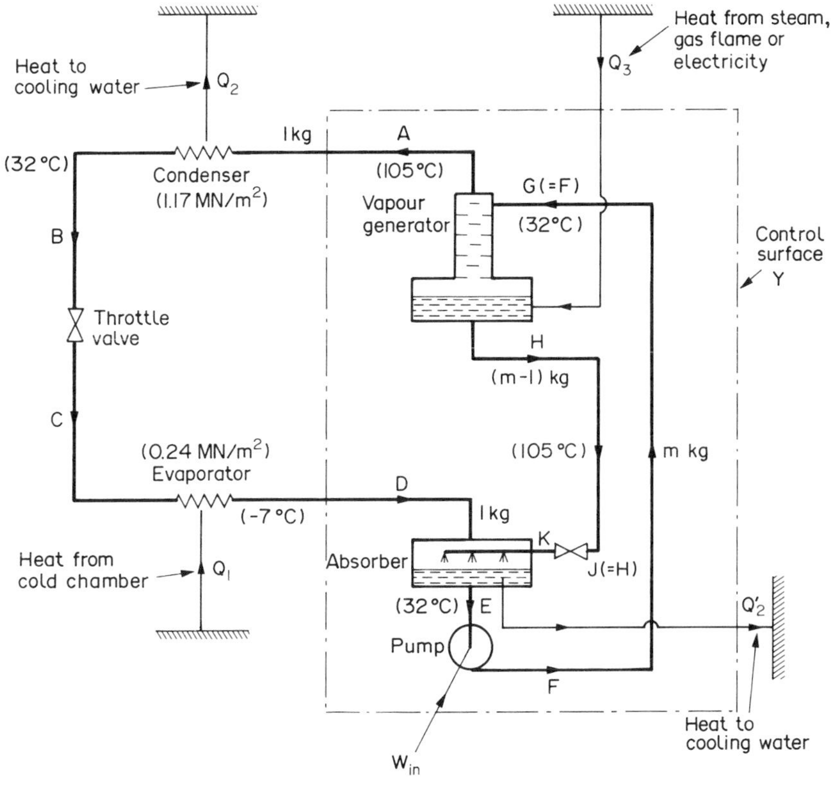

Fig. P.10.1

The energy balance for the evaporator gives:

$$Q_1 = [0.303 \times (-200) + 0.697 \times 1270] - 1 \times 100$$
$$= [-60.6 + 885.2] - 100$$
$$= 824.6 - 100 = \underline{725} \text{ kJ}$$

(d) Absorber

The energy balance for the absorber gives:

$$Q_2' = [824.6 + 6.23 \times 280] - [7.23 \times (-110)]$$
$$= 824.6 + 1744.4 + 795.3 = \underline{3364} \text{ kJ}$$

[CHECK: Neglecting pump work, $Q_1 + Q_3 = Q_2 + Q_2'$

$(725 + 4109) = (1470 + 3364)$]

(e) Coefficient of performance

$$CP \approx Q_1/Q_3 = 725/4109 = \underline{0.176} \qquad (10.3)$$

(f) Ideal plant

$$T_1 = (273.15 - 7) = 266.15 \text{ K}$$
$$T_2 = (273.15 + 32) = 305.15 \text{ K}$$
$$T_3 = (273.15 + 105) = 378.15 \text{ K}$$
$$T_2/T_1 = 1.1465$$
$$T_2/T_3 = 0.8070$$

From eqn. (10.6) in Ref. 1:

$$\text{Ideal CP} = \frac{[1 - (T_2/T_3)]}{[(T_2/T_1) - 1)]} = \frac{0.1930}{0.1465} = \underline{1.32}$$

$$\text{Reversed-Carnot CP} = \frac{T_1}{T_2 - T_1} = \frac{266.15}{39} = \underline{6.82}$$

10.2 Derive eqn. (10.8) in Ref. 1.

Solution

With reference to Fig. 10.2 in Ref. 1, it is required to show that the overall coefficient of performance, C_0, of the binary vapour-compression cycle is given by:

$$\left(1 + \frac{1}{C_0}\right) = \left(1 + \frac{1}{C_1}\right)\left(1 + \frac{1}{C_2}\right). \qquad (10.8)$$

By definition, the coefficient of performance of a refrigeration cycle is given by:

$$CP \equiv \frac{Q_{in}}{W_{in}}$$

$$\therefore \left(1 + \frac{1}{CP}\right) = \frac{Q_{out}}{Q_{in}}$$

Hence:

For the binary cycle, $\left(1 + \frac{1}{C_0}\right) = \frac{Q_3}{Q_1}$ (1)

For cycle 1, $\left(1 + \frac{1}{C_1}\right) = \frac{Q_2}{Q_1}$ (2)

For cycle 2, $\left(1 + \frac{1}{C_2}\right) = \frac{Q_3}{Q_2}$ (3)

Hence, from eqns. (1), (2) and (3):

$$\left(1 + \frac{1}{C_0}\right) = \left(1 + \frac{1}{C_1}\right)\left(1 + \frac{1}{C_2}\right)$$

10.3 In the "dry-ice" process illustrated in Figs. 10.5 and 10.6, gaseous CO_2 is drawn in at A at a pressure of 1 bar and a temperature of 20 °C, and solid carbon dioxide is withdrawn from the snow chamber, in which the solid and vapour are in equilibrium at a pressure of 1 bar. The pressures at exit from each of the compressor stages are respectively 6, 20 and 70 bar. In each case, dry saturated vapour enters the compressor stage and saturated liquid enters the throttle. Stray heat transfers with the environment may be neglected.

Calculate, per kg of solid CO_2 produced, (a) the values of m_1, m_2 and m_3; (b) the total work input; (c) the theoretical work input for a completely reversible process operating between the same initial and final states of the CO_2 in the presence of an environment at 20 °C; (d) the rational efficiency of the actual process; (e) the coefficient of performance of the plant if, instead of producing dry ice, it is used as a cyclic refrigerating plant with the supply of make-up gas from A shut off, and the solid-vapour mixture at M is turned into saturated vapour at B as a result of the transfer of heat from a refrigerating chamber to the snow chamber.

Check that the coefficient of performance of the cyclic plant calculated in (e) above is also given by the expression

$$\left(1 + \frac{1}{C_0}\right) = \left(1 + \frac{1}{C_1}\right)\left(1 + \frac{1}{C_2}\right)\left(1 + \frac{1}{C_3}\right),$$

where C_1, C_2 and C_3 are the respective coefficients of performance of the three vapour-compression refrigerating cycles BCDLMB, DEFJKD and FGHIF, so that its performance is equivalent to that of a ternary vapour-compression cycle. Explain why this result holds.

The relevant properties of CO_2 are as follows:

$h_A = 803.5$, $h_C = 804.0$, $h_E = 782.6$, $h_G = 789.1$ kJ/kg; $s_A = 4.841$ kJ/kg K

Pressure bar	Saturated Solid h kJ/kg	Saturated Solid s kJ/kg K	Sat. liquid h kJ/kg	Sat. vapour h kJ/kg
1	151.5	1.568	-	723.0
6	-	-	386.5	729.6
20	-	-	452.0	735.0
70	-	-	592.5	-

From Thermophysical Properties of Carbon Dioxide, Vukalovich, M.P. and Altunin, V.V.

Solution

Using the notation of Figs. P.10.3(a) and P.10.3(b):

(a) m_1, m_2 and m_3

Applying the steady-flow energy equation successively to the snow chamber and the two flash tanks gives:

$$m_1 = \frac{h_A - h_N}{h_B - h_L} = \frac{803.5 - 151.5}{723.0 - 386.5} = \frac{652.0}{336.5} = \underline{1.9376} \text{ kg}$$

$$\frac{m_2}{m_1} = \frac{h_C - h_L}{h_D - h_J} = \frac{804.0 - 386.5}{729.6 - 452.0} = \frac{417.5}{277.6}$$

$$\therefore m_2 = 1.9376 \times \frac{417.5}{277.6} = \underline{2.9141} \text{ kg}$$

$$\frac{m_3}{m_2} = \frac{h_E - h_J}{h_F - h_H} = \frac{782.6 - 452.0}{735.0 - 592.5} = \frac{330.6}{142.5}$$

$$\therefore m_3 = 2.9141 \times \frac{330.6}{142.5} = \underline{6.7607} \text{ kg}$$

(b) Total work input

$$W_1 = m_1(h_C - h_B) = 1.9376(804.0 - 723.0) = 156.9 \text{ kJ}$$

$$W_2 = m_2(h_E - h_D) = 2.9141(782.6 - 729.6) = 154.4 \text{ kJ}$$

$$W_3 = m_3(h_G - h_F) = 6.7607(789.1 - 735.0) = \underline{365.8} \text{ kJ}$$

$\therefore$ Total work input = $\underline{\mathit{677.1}}$ kJ/kg

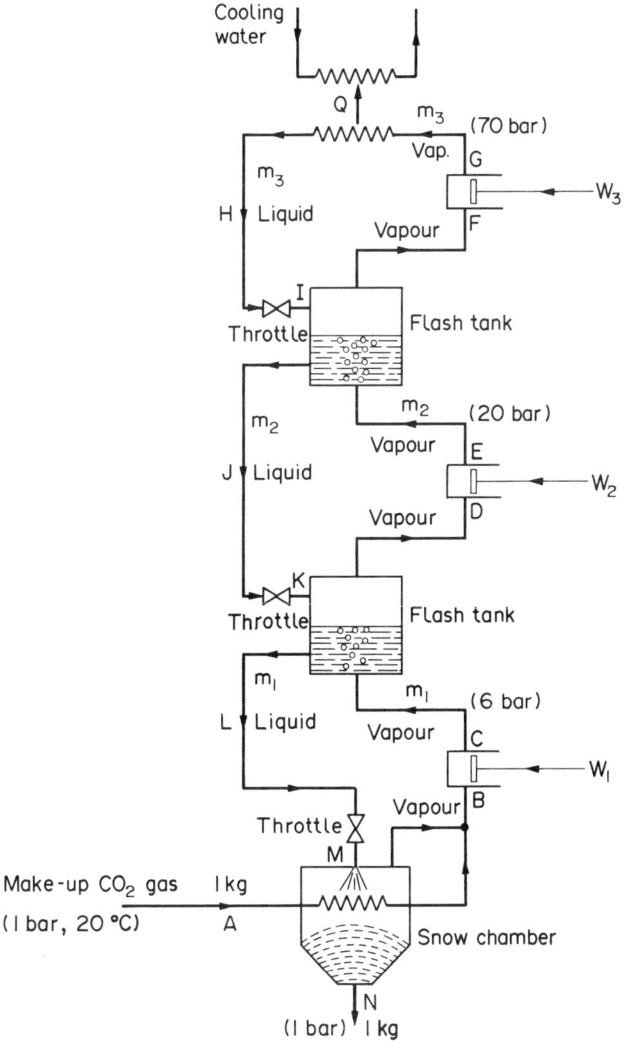

Fig. P.10.3(a)

(c) Theoretical work input

$$\left[(W_{in})_{REV}\right]_A^N = (b_N - b_A), \text{ with } t_0 = 20\ °C,\ T_0 = 293.15\ K$$

$$= (h_N - h_A) - T_0(s_N - s_A)$$

$$= (151.5 - 803.5) - 293.15(1.568 - 4.841)$$

$$= -652.0 + 959.5$$

$$= \underline{307.5}\ \text{kJ/kg}$$

(d) Rational efficiency

$$\eta_R \equiv \frac{(W_{in})_{REV}}{\text{Actual } W_{in}} = \frac{307.5}{677.1} \times 100 = \underline{45.4}\ \%$$

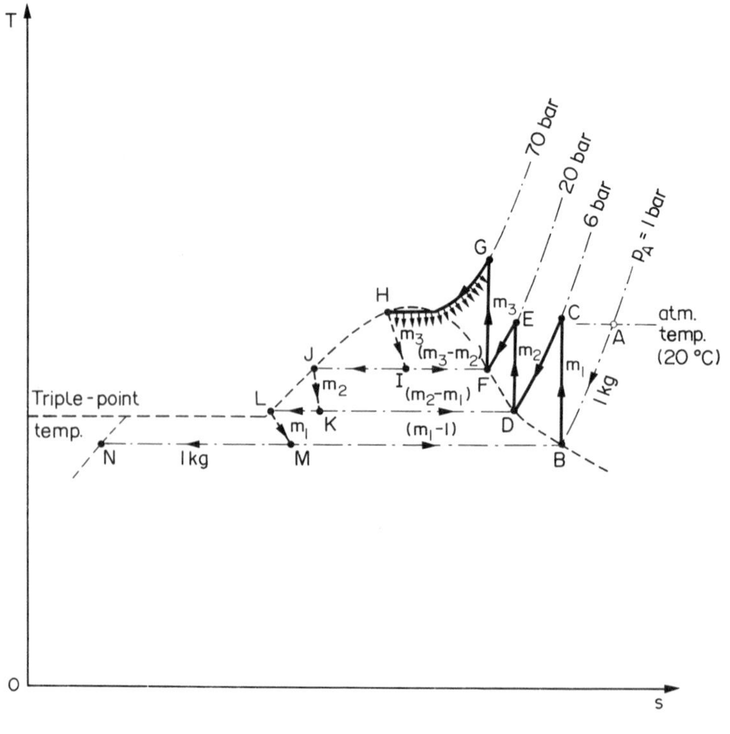

Fig. P.10.3(b)

(e) Coefficient of performance as refrigerating plant

As a cyclic refrigerating plant under the specified conditions:

$$Q_{out} = m_3(h_G - h_H) = 6.7607(789.1 - 592.5) = 1329.2 \text{ kJ}$$

$$Q_{in} = m_1(h_B - h_L) = 1.9376(723.0 - 386.5) = 652.0 \text{ kJ}$$

$$W_{in} = (Q_{out} - Q_{in}) = 677.2 \text{ kJ}$$

$$\text{Overall CP} \equiv \frac{Q_{in}}{W_{in}} = \frac{652.0}{677.2} = \underline{0.963}$$

Check against the given formula:

$$C_1 \equiv CP_1 = \frac{h_B - h_L}{h_C - h_B} = \frac{336.5}{81.0} = 4.154$$

$$C_2 \equiv CP_2 = \frac{h_D - h_J}{h_E - h_D} = \frac{277.6}{53.0} = 5.238$$

$$C_3 \equiv CP_3 = \frac{h_F - h_H}{h_G - h_F} = \frac{142.5}{54.1} = 2.634$$

$$\left(1 + \frac{1}{C_0}\right) = \left(1 + \frac{1}{C_1}\right)\left(1 + \frac{1}{C_2}\right)\left(1 + \frac{1}{C_3}\right)$$

$$= \frac{5.154}{4.154} \times \frac{6.238}{5.238} \times \frac{3.634}{2.634} = 2.0386$$

$$\therefore C_0 = \frac{1}{1.0386} = \underline{0.963} \text{ (as before)}$$

This result holds because the energy conservation equation for the direct-contact process in each flash tank is the same as would apply to an ideal heat exchanger which replaced the flash tank. The plant would then constitute a ternary cascade refrigerating cycle similar to the binary cycle of Fig. 10.2 of Ref. 1, to which equation (10.8) can readily be shown to apply.

10.4 In a gas-liquefaction plant, gas at the environment temperature T_0 and 1 atm pressure is first compressed to N atm and cooled in an after-cooler to T_0. The gas then enters a refrigerating plant in which it is wholly liquefied, leaving as saturated liquid at 1 atm. The compressor has an isothermal efficiency of η_C. The rational efficiency of the refrigerating process, excluding the compressor and aftercooler, is η_R.

The gas may be treated as a perfect gas at all pressures when at temperature T_0, and also at all temperatures when gaseous at 1 atm pressure. At 1 atm pressure, the boiling point is T_0/n and the specific enthalpy of evaporation is L.

Derive an expression for the total work input to the plant per unit mass of gas liquefied.

Solution

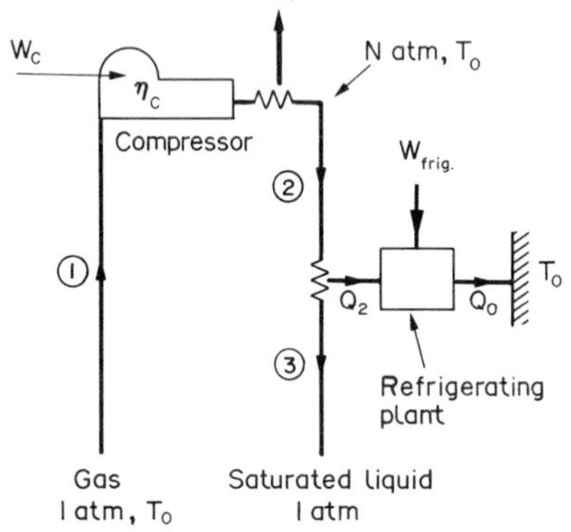

Fig. P.10.4(a)

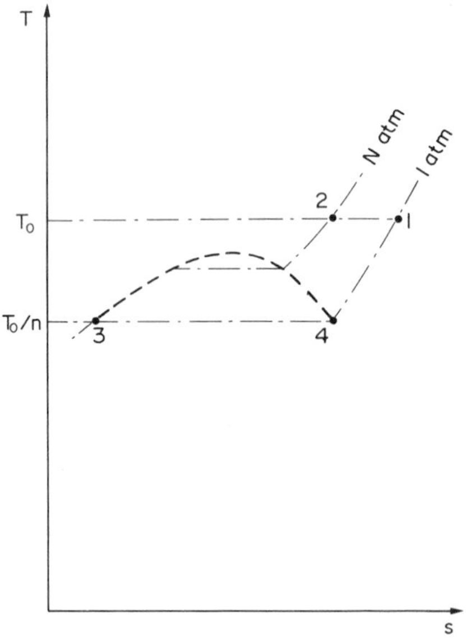

Fig. P.10.4(b)

Compressor

For reversible, isothermal compression at T_0 from 1 atm to N atm:

$$(W_{in})_{REV} = \int_1^N v\,dp = R\,T_0 \int_1^N (dp/p) = R\,T_0 \ln N$$

$$\therefore W_C = \frac{R\,T_0}{\eta_c} \ln N \qquad (1)$$

Refrigerating process

For a fully reversible, steady-flow refrigerating process in which heat is exchanged reversibly with the environment at T_0, the work input would be given by

$$(W_{frig.})_{REV} = (b_3 - b_2) = (h_3 - h_2) - T_0(s_3 - s_2) \qquad (2)$$

For a perfect gas at constant temperature, h is constant. Hence, with reference to Fig. P.10.4(b):

$$(h_2 - h_3) = (h_2 - h_1) + (h_1 - h_4) + (h_4 - h_3)$$

$$= 0 + c_p\left[T_0 - \frac{T_0}{n}\right] + L$$

$$= c_p\,T_0\left(1 - \frac{1}{n}\right) + L \qquad (3)$$

Now, $T\,ds = dh - v\,dp$

$$(s_2 - s_3) = (s_2 - s_1) + (s_1 - s_4) + (s_4 - s_3)$$

$$(s_2 - s_1) = -\int_1^2 \frac{v}{T}\,dp = -R\int_1^N \frac{dp}{p} = -R \ln N$$

$$(s_1 - s_4) = c_p\int_{T_0/n}^{T_0} \frac{dT}{T} = c_p\left[\ln T_0 - \ln \frac{T_0}{n}\right] = c_p \ln n$$

$$(s_4 - s_3) = \frac{L}{T_0/n} = \frac{L\,n}{T_0}$$

$$\therefore (s_2 - s_3) = -R \ln N + c_p \ln n + \frac{L\,n}{T_0} \qquad (4)$$

But the actual work input to the refrigerating process is given by

$$W_{frig.} = \frac{(W_{frig.})_{REV}}{\eta_R}, \qquad (5)$$

where η_R is the rational efficiency of the refrigerating process. Hence, from equations (1) to (5), the total work input to the plant is given by

$$\frac{c_p\, T_0}{\eta_R}\left(ln\ n + \frac{1}{n} - 1\right) + \frac{L}{\eta_R}(n - 1) - \left(\frac{1}{\eta_R} - \frac{1}{\eta_C}\right) R\, T_0\ ln\ N$$

10.5 Prove eqn. (10.14), and show that μ_h is zero for a perfect gas. Noting that, at the critical point, $p = p_c$, $T = T_c$, $v = v_c$ and

$$\left(\frac{\partial p}{\partial v}\right)_T = \left(\frac{\partial^2 p}{\partial v^2}\right)_T = 0,$$

express van der Waals' equation of state (Section 10.9) in terms of the reduced coordinates, $p_R \equiv p/p_c$, $T_R \equiv T/T_c$ and $v_R \equiv v/v_c$. Show that points on the inversion line (for which the isenthalpic Joule-Thomson coefficient μ_h is zero) are given by the relations

$$T_R = \frac{3(3v_R - 1)^2}{4v_R^2} \quad \text{and} \quad p_R = \frac{9(2v_R - 1)}{v_R^2}.$$

Sketch the inversion line on the pressure-volume and pressure-temperature planes.

Solution

Equation (10.14) in Ref. 1 is:

$$- c_p\, \mu_h = \mu_T = v - T\left(\frac{\partial v}{\partial T}\right)_p,$$

where $\mu_h \equiv \left(\frac{\partial T}{\partial p}\right)_h \equiv$ isenthalpic Joule-Thomson coefficient,

$\mu_T \equiv \left(\frac{\partial h}{\partial p}\right)_T \equiv$ isothermal Joule-Thomson coefficient.

To prove equation (10.14):

Let $h = h(T,p)$

Then $$\delta h = \left(\frac{\partial h}{\partial T}\right)_p \delta T + \left(\frac{\partial h}{\partial p}\right)_T \delta p$$

∴ For any small step along the (h,T,p) three-dimensional surface,

$$\frac{\delta h}{\delta T} = \left(\frac{\partial h}{\partial T}\right)_p + \left(\frac{\partial h}{\partial p}\right)_T \frac{\delta p}{\delta T}$$

Hence, for a step at constant h along this surface:

$$\left(\frac{\delta h}{\delta T}\right)_h = \left(\frac{\partial h}{\partial T}\right)_p + \left(\frac{\partial h}{\partial p}\right)_T \left(\frac{\delta p}{\delta T}\right)_h$$

But $\left(\frac{\delta h}{\delta T}\right)_h = 0$ and $\left(\frac{\partial h}{\partial T}\right)_p \equiv c_p$

$$\therefore\ 0 = c_p + \mu_T(1/\mu_h)$$

$$- c_p\ \mu_h = \mu_T \tag{1}$$

Now $\delta h = v\ \delta p + T\ \delta s$

$$\therefore \left(\frac{\partial h}{\partial p}\right)_T = v + T\left(\frac{\partial s}{\partial p}\right)_T \tag{2}$$

But we have the Maxwell relation* (Ref. 3, p. 278)

$$\left(\frac{\partial s}{\partial p}\right)_T = -\left(\frac{\partial v}{\partial T}\right)_p \tag{3}$$

Whence, from (2) and (3):

$$\mu_T = v - T\left(\frac{\partial v}{\partial T}\right)_p \tag{4}$$

Thus, from (1) and (4):

$$\underline{\underline{-\,c_p\ \mu_h = \mu_T = v - T\left(\frac{\partial v}{\partial T}\right)_p}} \tag{10.14}$$

*<u>To prove the above Maxwell relation</u>, we have:

By definition of the Gibbs function:

$$g \equiv h - Ts$$

$$\therefore\ \delta g = \delta h - T\ \delta s - s\ \delta T = v\ \delta p - s\ \delta T$$

Whence $s = -\left(\frac{\partial g}{\partial T}\right)_p$ and $v = \left(\frac{\partial g}{\partial p}\right)_T$

$$\therefore \left(\frac{\partial s}{\partial p}\right)_T = -\frac{\partial^2 g}{\partial p\ \partial T} = -\frac{\partial^2 g}{\partial T\ \partial p} = -\left(\frac{\partial v}{\partial T}\right)_p$$

<u>For a perfect gas</u>, $v = \frac{RT}{p}$

$$\therefore \left(\frac{\partial v}{\partial T}\right)_p = \frac{R}{p}$$

Whence $\left[v - T\left(\frac{\partial v}{\partial T}\right)_p\right] = v - \frac{RT}{p} = 0$

Hence, from eqn. (10.14), $\underline{\underline{\mu_h = 0}}$

The van der Waals equation of state is

$$\left(p + \frac{a}{v^2}\right)(v - b) = RT$$

$$p = \frac{RT}{v - b} - \frac{a}{v^2} \tag{5}$$

∴ At the critical point, when $p = p_c$, $v = v_c$ and $T = T_c$,

$$\left(\frac{\partial p}{\partial v}\right)_T = -\frac{RT_c}{(v_c - b)^2} + \frac{2a}{v_c^3} = 0 \qquad \therefore \frac{RT_c}{(v_c - b)^2} = \frac{2a}{v_c^3} \tag{6}$$

$$\left(\frac{\partial^2 p}{\partial v^2}\right)_T = \frac{2RT_c}{(v_c - b)^3} - \frac{6a}{v_c^4} = 0 \qquad \therefore \frac{RT_c}{(v_c - b)^3} = \frac{3a}{v_c^4} \tag{7}$$

Dividing (6) by (7) gives:

$$(v_c - b) = \frac{2}{3} v_c \qquad \therefore v_c = 3b \tag{8}$$

Substituting (8) in (6) gives:

$$RT_c = \frac{2a(v_c - b)^2}{R\, v_c^3} = \frac{2a \times (2b)^2}{27b^3} \qquad \therefore RT_c = \frac{8a}{27b} \tag{9}$$

$$\text{Now } p_c = \frac{RT_c}{v_c - b} - \frac{a}{v_c^2} \tag{10}$$

Whence, substituting (8) and (9) in (10), $p_c = \frac{a}{27b^2}$ (11)

In terms of reduced coordinates:

$$p = p_R p_c, \qquad v = v_R v_c, \qquad T = T_R T_c$$

Substituting these in (5) gives:

$$p_R p_c = \frac{RT_R T_c}{(v_R v_c - b)} - \frac{a}{v_R^2 v_c^2} \tag{12}$$

Whence, from (8), (9), (11) and (12):

$$p_R = \frac{8T_R}{3v_R - 1} - \frac{3}{v_R^2} \tag{13}$$

On the inversion line:

$$\mu_h \equiv \left(\frac{\partial T}{\partial p}\right)_h = 0,$$

so that, from eqn. (10.14):

$$\mu_T \equiv \left(\frac{\partial h}{\partial p}\right)_T = v - T\left(\frac{\partial v}{\partial T}\right)_p = 0$$

Thus, on the inversion line, in terms of reduced coordinates:

$$\left(\frac{\partial T_R}{\partial v_R}\right)_{p_R} = \frac{T_R}{v_R} \tag{14}$$

At constant p_R, we have from (13):

$$\frac{8}{3v_R - 1}\,\delta T_R - \frac{24T_R}{(3v_R - 1)^2}\,\delta v_R + \frac{6}{v_R^3}\,\delta v_R = 0 \tag{15}$$

∴ From (14) and (15):

$$\left(\frac{\partial T_R}{\partial v_R}\right)_{p_R} = \frac{3T_R}{3v_R - 1} - \frac{3(3v_R - 1)}{4v_R^3} = \frac{T_R}{v_R}$$

Hence, points on the inversion line are given by:

$$T_R = \frac{3(3v_R - 1)^2}{4v_R^2} \tag{16}$$

and, from (13) and (16):

$$p_R = \frac{9(2v_R - 1)}{v_R^2} \tag{17}$$

Thus, on the p_R, v_R and p_R, T_R planes, the inversion line appears as shown below.

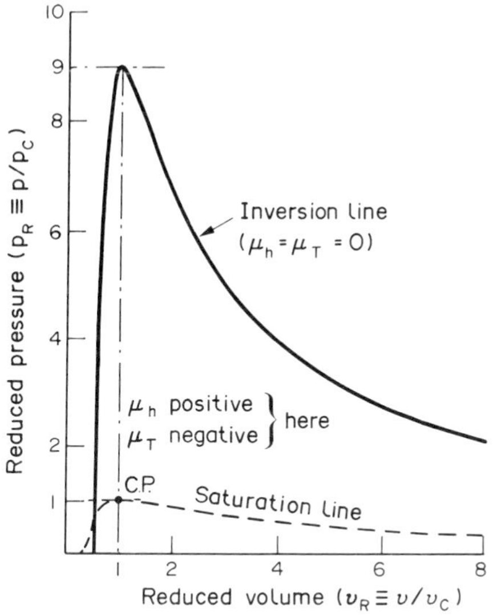

Fig. P.10.5(a)

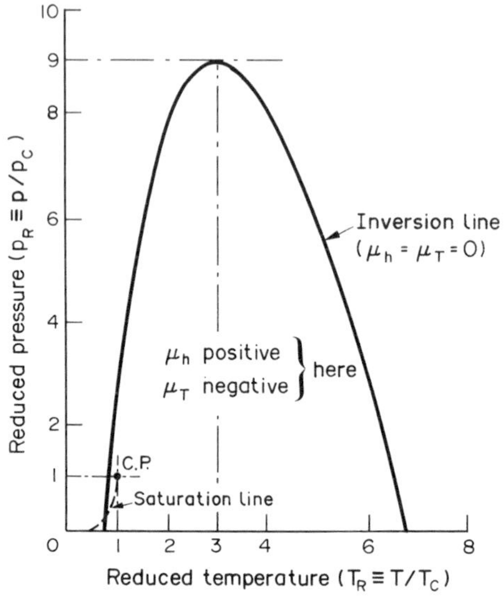

Fig. P.10.5(b)

Problems 10.6 to 10.12 relate to processes for the production of saturated liquid air at a pressure of 1 bar from air at 1 bar and 290 K, which is also the temperature of the environment.

10.6 Calculate the ideal work input for a completely reversible steady-flow process operating between the end states specified above, with the environment at the same temperature. Express the result both in MJ/kg and kW h/litre of liquid air produced.

Solution

State 1 = Air at 1 bar and 290 K
State 2 = Saturated liquid air at 1 bar
(both states: 1 bar ≡ 0.1 MN/m²)

T_0 = environment temperature = 290 K

From Tables 17, 18 and 19 of Ref. 2:

State	$\frac{h}{\text{kJ/kg}}$	$\frac{s}{\text{kJ/kg K}}$
1	417.8	3.858 (Taking s α ln T)
2	0	0

$$(w_{in})_{REV} = (b_2 - b_1)$$

$$= (h_2 - h_1) - T_0(s_2 - s_1)$$

$$= -417.8 + 290 \times 3.858$$

$$= 701.0 \text{ kJ/kg} = \underline{0.701} \text{ MJ/kg}$$

Specific volume of saturated liquid air at 1 bar = 0.00114 m^3/kg

= 1.14 litre/kg

$$(w_{in})_{REV} = \frac{701.0}{3600 \times 1.14} = \underline{0.171} \text{ kW h/litre}$$

10.7 In a simple Linde process, the high-pressure air entering the regenerative heat exchanger and the low-pressure air leaving it are at 290 K. For a range of compressor delivery pressure from 250 to 500 bar, draw a graph showing the yield y plotted against the delivery pressure, and so determine the optimum pressure for maximum yield. At this pressure calculate the temperature of the air entering the throttle.

For the same range of compressor delivery pressure plot the work input per unit yield when the compression process is isothermal and reversible. Thence estimate the optimum pressure for maximum efficiency (i.e. minimum work input per unit yield), and the value of this work input, expressed both in MJ/kg and kW h/litre. Calculate the corresponding rational efficiency of the process.

Solution

The numbering of the state points in Fig. P.10.7(a) is the same as that in Fig. 10.8(b) of Ref. 1.

From Problem 10.6:

$$h_1 = 417.8 \text{ kJ/kg}$$

$$s_1 = 3.858 \text{ kJ/kg K}$$

From eqn. (10.15) in Ref. 1, when stray "heat leak" is zero, the yield y is given by

$$y = \frac{h_1 - h_5}{h_1 - h_8}. \quad \text{(Here, } h_8 = 0 \text{ kJ/kg)} \qquad (1)$$

For reversible, isothermal compression from state 1 to state 5:

$$W_{in} = (b_5 - b_1) = T_0(s_1 - s_5) - (h_1 - h_5) \qquad (2)$$

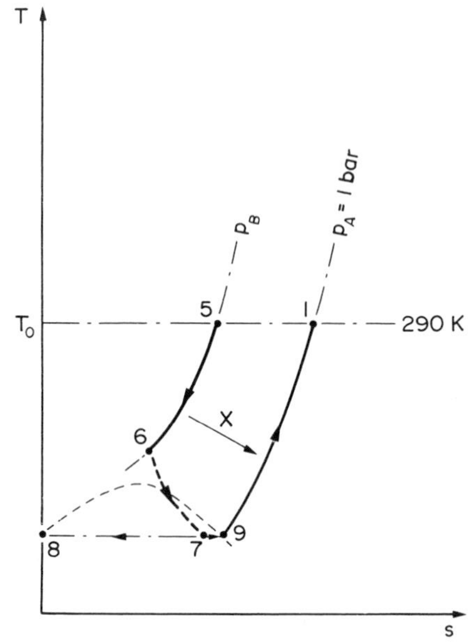

Fig. P.10.7(a)

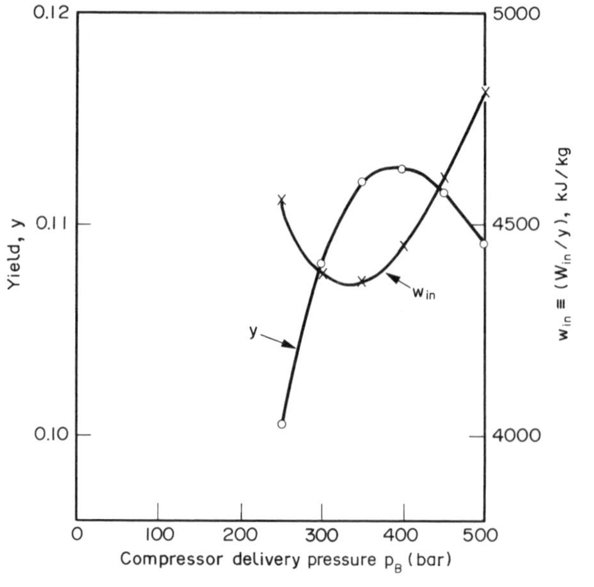

Fig. P.10.7(b)

Whence, from eqns. (1) and (2), noting that 1 bar ≡ 0.1 MN/m^2:

p_B	h_5	$(h_1 - h_5)$	y	s_5^*	$(s_1 - s_5)$	$T_0(s_1 - s_5)$	W_{in}	$w_{in} = W_{in}/y$
bar	kJ/kg			kJ/kg K			kJ/kg	
250	375.8	42.0	0.1005	2.134	1.724	500.0	458.0	4557
300	372.6	45.2	0.1082	2.067	1.791	519.4	474.2	4383
350	371.0	46.8	0.1120	2.012	1.846	535.3	488.5	4362
400	370.7	47.1	0.1127	1.965	1.893	549.0	501.9	4453
450	371.2	46.6	0.1115	1.925	1.933	560.6	514.0	4610
500	372.2	45.6	0.1091	1.889	1.969	571.0	525.4	4816

*Taking $s \propto \ln T$

Hence, from Fig. P.10.7(b), the optimum pressures are:

For maximum yield, y : About *400* bar

For maximum efficiency*: About *335* bar

*Minimum w_{in} ($\equiv W_{in}/y$)

At maximum yield (400 bar)

Yield $y = 0.1127$

$$h_6 = h_7 = y\,h_8 + (1 - y)h_9 = (1 - y)h_9, \text{ since } h_8 = 0$$

$$\therefore h_6 = 0.8873 \times 205.3 = 182.2 \text{ kJ/kg}$$

Hence, from Table 18 in Ref. 2:

$$t_6 = 160 + \left(\frac{182.2 - 170.5}{187.7 - 170.5}\right) \times 10 \approx \underline{167} \text{ K}$$

At maximum efficiency (335 bar)

From Fig. P.10.7(b):

$$\text{Minimum } w_{in} = 4360 \text{ kJ/kg} = \underline{4.36} \text{ MJ/kg}$$

$$= \frac{4360}{3600 \times 1.14} = \underline{1.06} \text{ kW h/litre}$$

$$\text{Rational efficiency, } \eta_R = \frac{(w_{in})_{REV} \text{ from Problem 10.6}}{w_{in}}$$

$$= \frac{0.701}{4.36} \times 100 = \underline{16.1}\ \%$$

10.8 In a simple Linde process, the high-pressure air entering the regenerative heat exchanger and the low-pressure air leaving it are at 290 K. The pressure at compressor delivery is 200 bar and the compression process is isothermal and reversible.

Calculate (a) the yield; (b) the temperature of the air entering the throttle; (c) the work input per unit yield, expressed both in MJ/kg and kW h/litre; (d) the rational efficiency of the process.

Evaluate the extra work input due to irreversibility arising from (i) the heat transfer process in the regenerative heat exchanger, (ii) the throttling process, expressing these quantities as a percentage of the actual work input.

If, instead of being used for air liquefaction, the plant were used as a cyclic refrigerating plant, what would be its coefficient of performance? Express this as a percentage of the coefficient of performance of a reversed-Carnot refrigerating plant operating between the same extremes of temperature. Take the saturation temperature of air at 1 bar to be 81.7 K, the dew point of the vapour at that pressure.

Solution

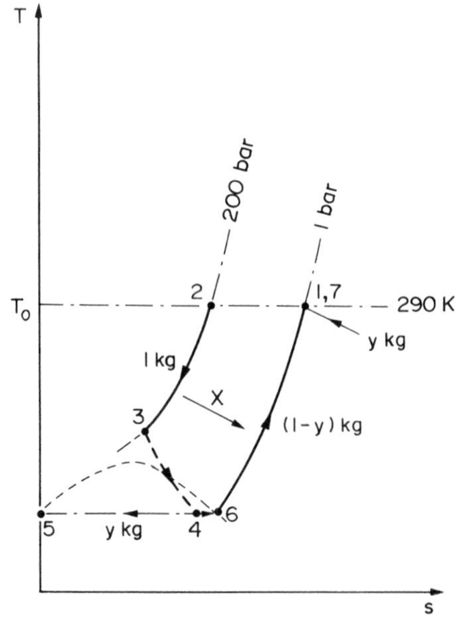

Fig. P.10.8(a)

State	$\frac{h}{kJ/kg}$	$\frac{s}{kJ/kg\ K}$
1,7	417.8	3.858
2	380.6	2.219
5	0	0
6	205.3	2.559

(a) The yield, y

The energy conservation equation for control volume Y gives:

$$\text{Yield, } y = \frac{h_7 - h_2}{h_7 - h_5}$$

$$\therefore\ y = \frac{37.2}{417.8} = \underline{0.0890}$$

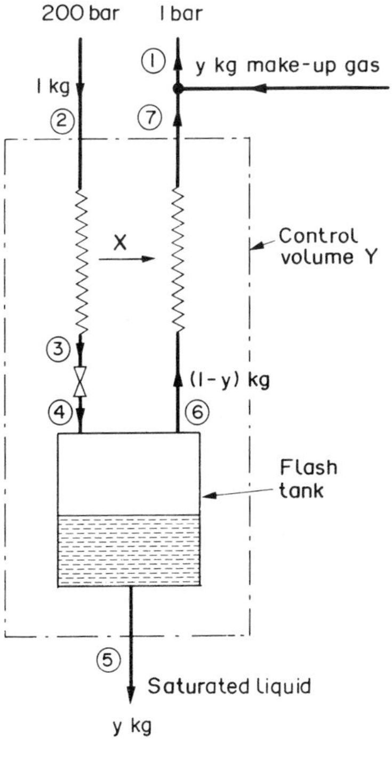

Fig. P.10.8(b)

(b) Calculation of T_3

The energy-conservation equation for the flash tank gives:

$$y = \frac{h_6 - h_4}{h_6 - h_5}$$

$$\therefore (h_6 - h_4) = 0.0890 \times 205.3 = 18.3 \text{ kJ/kg}$$

$$\therefore h_4 = 205.3 - 18.3 = 187.0 \text{ kJ/kg}$$

But $h_3 = h_4$

Hence, from Table 18 of Ref. 2, at p = 200 bar, h = 187.0 kJ/kg:

$$T_3 \approx \underline{170} \text{ K}$$

(c) Work input per unit yield

For reversible, isothermal compression at T_0 = 290 K:

$$W_{in} = (b_2 - b_1) = T_0(s_1 - s_2) - (h_1 - h_2)$$

$$= 290 \times 1.639 - 37.2 = 438.1 \text{ kJ/kg}$$

$$\therefore w_{in} \equiv \frac{W_{in}}{y} = \frac{438.1}{0.0890} \times 10^{-3} = \underline{4.92} \text{ MJ/kg}$$

$$= \frac{4.92 \times 10^3}{3600 \times 1.14} = \underline{1.20} \text{ kW h/litre}$$

(d) Rational efficiency

$$\text{Rational efficiency} = \frac{(w_{in})_{REV} \text{ from Problem 10.6}}{w_{in}}$$

$$= \frac{0.701}{4.92} \times 100 = \underline{14.2} \text{ \%}$$

(i) Heat transfer process in heat exchanger X

Extra work input due to irreversibility = $T_0 \, \Delta S_C$,

where ΔS_C is the entropy creation in the process.

$$\Delta S_C = (1 - y)(s_7 - s_6) - (s_2 - s_3)$$

At state 3, p = 200 bar, T_3 = 170 K

$$\therefore s_3 = 1.339 \text{ kJ/kg K}$$

$$(1 - y)(s_7 - s_6) = 0.9110(3.858 - 2.559)$$

$$= 1.183 \text{ kJ/kg K}$$

$$(s_2 - s_3) = (2.219 - 1.339)$$

$$= 0.880 \text{ kJ/kg K}$$

$$\therefore \Delta S_C = 1.183 - 0.880 = 0.303 \text{ kJ/kg K}$$

Per kg of liquid air:

$$\text{Extra work input} = \frac{T_0 \, \Delta S_C}{y} = \frac{290 \times 0.303}{0.0890} \times 10^{-3} = 0.987 \text{ MJ/kg}$$

$$\underline{\text{Percentage of actual work input}} = \frac{0.987}{4.92} \times 100 = \underline{20.1} \text{ \%}$$

(ii)   <u>Throttling process</u>

$$\Delta S_C = 1 \times (s_4 - s_3)$$

$$s_4 = (1 - y)s_6, \text{ since } s_5 = 0$$

$$= 0.9110 \times 2.559 = 2.331 \text{ kJ/kg K}$$

$$\therefore \Delta S_C = 2.331 - 1.339 = 0.992 \text{ kJ/kg K}$$

Per kg of liquid air:

$$\text{Extra work input} = \frac{T_0 \, \Delta S_C}{y} = \frac{290 \times 0.992}{0.0890} \times 10^{-3} = 3.232 \text{ MJ/kg}$$

$$\underline{\text{Percentage of actual work input}} = \frac{3.232}{4.92} \times 100 = \underline{65.7} \text{ \%}$$

<u>Check on the rational efficiency</u>

Total extra work input = 20.1 + 65.7   = 85.8 %

$$\therefore \text{ Rational efficiency, } \eta_R = (100 - 85.8) = \underline{14.2} \text{ \% (as before)}$$

<u>When operating as a refrigerating plant</u> and returning evaporated air to the compressor at $T_7 = T_0 = 290$ K:

$$(h_2 - h_3) = (h_7 - h_6)$$

$$\therefore h_3 = 380.6 + 205.3 - 417.8 = 168.1 \text{ kJ/kg}$$

$$\text{Coefficient of performance, CP} = \frac{h_6 - h_4}{W_{in}}$$

For the throttle, $h_4 = h_3 = 168.1$ kJ/kg

$$\therefore \text{CP} = \frac{205.3 - 168.1}{438.1} = \underline{0.0849}$$

$$\text{Reversed-Carnot CP} = \frac{T_5}{T_1 - T_5} = \frac{81.7}{290 - 81.7} = 0.3922$$

$$\frac{\text{CP}}{\text{Reversed-Carnot CP}} = \frac{0.0849}{0.3922} \times 100 = \underline{21.6}\ \%$$

10.9 In a dual-pressure Linde process (Fig. 10.10), the intermediate pressure p_B is supercritical. The compression process is isothermal and reversible, and the operating conditions are as follows: p_B = 50 bar, p_C = 200 bar, $T_{11} = T_7 = T_2 = T_1$ and $y_B = 0.2$.

Calculate (a) the yield y; (b) the values of T_3 and T_4; (c) the wetness y_A of the mixture leaving the low-pressure throttle; (d) the work input per unit yield, expressed both in MJ/kg and kW h/litre; (e) the rational efficiency of the process.

Evaluate the extra work input due to irreversibility arising from (i) the heat transfer process in the regenerative heat exchangers, (ii) the HP throttling process, (iii) the LP throttling process, expressing these quantities as a percentage of the actual work input.

Note - *Here (a) and (b) have been interchanged compared with the original text of Ref. 1, in which they were listed as (b) and (a) respectively. This was done because the yield y is directly calculable first.*

Solution (Dual-pressure Linde process-Supercritical intermediate pressure)

State	h kJ/kg	s kJ/kg K
1,11	417.8	3.858
2	380.6	2.219
7	407.0	2.704
9	0	0
10	205.3	2.559

Since p_B is supercritical, no liquid is formed in receiver B, so that the fluid remains single-phase from 2 to 5, and

$h_3 = h_4 = h_5 = h_6 = h_8$.

In flash tank A, a fraction y_A of y_B leaves as liquid, so that the yield y of the plant is given by

$$y = y_A y_B .$$

(a) The yield, y

The energy conservation equation for control volume Y gives [eqn. (10.22) of Ref. 1]:

$$h_2 = (1 - y_B)h_7 + (y_B - y)h_{11} + y\,h_9$$

$$380.6 = 0.8 \times 407.0 + (0.2 - y) \times 417.8$$

$$\therefore\ y = \frac{28.6}{417.8} = \underline{0.06845}$$

(b) Calculation of T_3 and T_4

It is first necessary to determine the value of h_3. Noting that $h_8 = h_3$, the energy conservation equation for flash tank A gives [eqn. (10.23) of Ref. 1]:

$$y_A = \frac{y}{y_B} = \frac{h_{10} - h_3}{h_{10} - h_9}$$

$$\frac{0.06845}{0.2} = \frac{205.3 - h_3}{205.3}$$

$$h_3 = 205.3 - 70.3 = 135.0 \text{ kJ/kg}$$

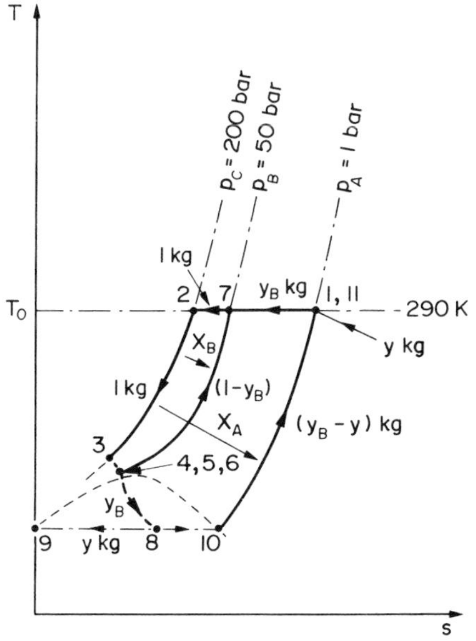

Fig. P.10.9(a)

∴ From Table 18 of Ref. 2, at p = 200 bar, h = 135.0 kJ/kg:

$$T_3 = 145 + \left(\frac{135.0 - 131.1}{142.7 - 131.1}\right) \times 5 = \underline{146.7} \text{ K}$$

$$h_4 = h_3 = 135.0 \text{ kJ/kg}$$

∴ From Table 18 of Ref. 2, at p = 50 bar, h = 135.0 kJ/kg

$$T_4 = 135 + \left(\frac{135.0 - 134.4}{165.4 - 134.4}\right) \times 5 = \underline{135.1} \text{ K}$$

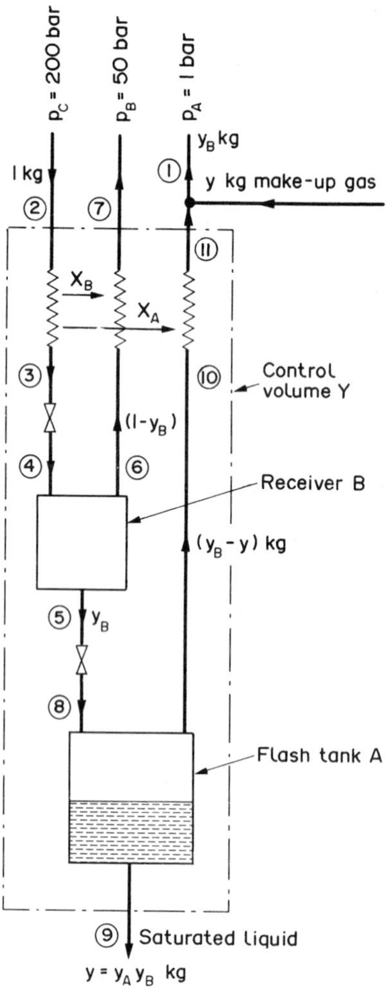

Fig. P.10.9(b)

(c) The wetness y_A at state point 8

$$y_A = \frac{y}{y_B} = \frac{0.06845}{0.2} = \underline{0.342}$$

(d) Work input per unit yield

For reversible, isothermal compression at T_0 = 290 K:

Compression from state 1 to state 7:

$$\text{Work input per kg} = (b_7 - b_1)$$

$$= T_0(s_1 - s_7) - (h_1 - h_7)$$

$$= 290 \times 1.154 - 10.8 = 323.9 \text{ kJ/kg}$$

$$\therefore [W_{in}]_1^7 = 323.9 \times y_B = 323.9 \times 0.2 = 64.8 \text{ kJ/kg}$$

Compression from state 7 to state 2:

$$[W_{in}]_7^2 = (b_2 - b_7)$$

$$= T_0(s_7 - s_2) - (h_7 - h_2)$$

$$= 290 \times 0.485 - 26.4 = 114.3 \text{ kJ/kg}$$

$$\therefore \text{Total } W_{in} = 64.8 + 114.3 = 179.1 \text{ kJ/kg delivered}$$

Work input per unit yield, $w_{in} \equiv W_{in}/y$

$$\therefore w_{in} = \frac{179.1}{0.06845} \times 10^{-3} = \underline{2.617} \text{ MJ/kg}$$

$$= \frac{2.617 \times 10^3}{3600 \times 1.14} = \underline{0.638} \text{ kW h/litre}$$

(e) Rational efficiency

$$\text{Rational efficiency} = \frac{(w_{in})_{REV} \text{ from Problem 10.6}}{w_{in}}$$

$$= \frac{0.701}{2.617} \times 100 = \underline{26.8}\ \%$$

(i) Heat transfer process in heat exchangers X_A and X_B

Extra work input due to irreversibility = $T_0\ \Delta S_C$

$$\Delta S_C = [(1 - y_B)(s_7 - s_6) + (y_B - y)(s_{11} - s_{10})] - 1 \times (s_2 - s_3)$$

State 3: $p_3 = 200$ bar, $T_3 = 146.7$ K

From Table 19 of Ref. 2 (taking $s \propto \ln T$):

$$s_3 = 1.0075 \text{ kJ/kg K}$$

State 6: $p_6 = 50$ bar, $T_6 = T_4 = 135.1$ K (since p_B is supercritical)

From Table 19 of Ref. 2, noting that $s \propto \ln T$:

$$s_6 = 1.1903 \text{ kJ/kg K}$$

$$(1 - y_B)(s_7 - s_6) = 0.8 \times 1.5137 = 1.2110 \text{ kJ/kg K}$$

$$(y_B - y)(s_{11} - s_{10}) = 0.13155 \times 1.299 = 0.1709 \text{ kJ/kg K}$$

$$1.3819 \text{ kJ/kg K}$$

$$1 \times (s_2 - s_3) = 2.219 - 1.0075 = 1.2115 \text{ kJ/kg K}$$

$$\therefore \Delta S_C = 0.1704 \text{ kJ/kg K}$$

Per kg of liquid air:

$$\text{Extra work input} = \frac{T_0 \, \Delta S_C}{y} = \frac{290 \times 0.1704}{0.06845} \times 10^{-3} = 0.722 \text{ MJ/kg}$$

$$\text{Percentage of actual work input} = \frac{0.722}{2.617} \times 100 = \underline{27.6}\ \%$$

(ii) HP throttling process

$$\Delta S_C = 1 \times (s_4 - s_3)$$

Since p_B is supercritical, $s_4 = s_6 = 1.1903$ kJ/kg K

$$\therefore \Delta S_C = (1.1903 - 1.0075) = 0.1828 \text{ kJ/kg K}$$

Per kg of liquid air:

$$\text{Extra work input} = \frac{T_0 \, \Delta S_C}{y} = \frac{290 \times 0.1828}{0.06845} \times 10^{-3} = 0.774 \text{ MJ/kg}$$

$$\text{Percentage of actual work input} = \frac{0.774}{2.617} \times 100 = \underline{29.6}\ \%$$

(iii) LP throttling process

$$\Delta S_C = y_B(s_8 - s_5)$$

Since p_B is supercritical, $s_5 = s_6 = 1.1903$ kJ/kg K

$s_8 = (1 - y_A)s_{10}$, since $s_9 = 0$

$= 0.658 \times 2.559 = 1.6838$ kJ/kg K

$\therefore \Delta s_C = 0.2 \times 0.4935 = 0.0987$ kJ/kg K

Per kg of liquid air:

$$\text{Extra work input} = \frac{T_0\ \Delta s_C}{y} = \frac{290 \times 0.0987}{0.06845} \times 10^{-3} = 0.418 \text{ MJ/kg}$$

$$\text{Percentage of actual work input} = \frac{0.418}{2.617} \times 100 = \underline{16.0}\ \%$$

Check on the rational efficiency

Total extra work input = 27.6 + 29.6 + 16.0 = 73.2 %

$\therefore$ Rational efficiency, η_R = (100 - 73.2) = *26.8* % (as before)

10.10 In a dual-pressure Linde process (Fig. 10.10), the intermediate pressure is subcritical. The compression process is isothermal and reversible, and the operating conditions are as follows: p_B = 30 bar, p_C = 200 bar, $T_{11} = T_7 = T_2 = T_1$.

Calculate (a) the wetness y_A of the mixture leaving the LP throttle; (b) the value of y_B for which the plant must be designed in order to satisfy the above conditions; (c) the yield y; (d) the value of T_3; (e) the work input per unit yield, expressed both in MJ/kg and kW h/litre; (f) the rational efficiency of the process.

Evaluate the extra work input due to irreversibility arising from (i) the heat transfer process in the regenerative heat exchangers, (ii) the HP throttling process, (iii) the LP throttling process, expressing these quantities as a percentage of the actual work input.

Solution (Dual-pressure Linde process - Subcritical intermediate pressure)

State	h kJ/kg	s kJ/kg K
1,11	417.8	3.858
2	380.6	2.219
5	116.8	1.077
6	199.0	1.723
7	411.5	2.865
9	0	0
10	205.3	2.559

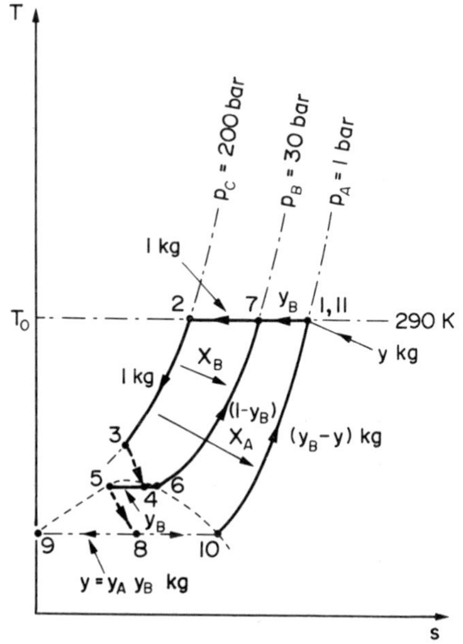

Fig. P.10.10(a)

Since p_B is subcritical, the fluid at exit from the HP throttle, at state 4, will be a liquid-vapour mixture. Hence liquid will appear in receiver B, which will now function as a flash tank in the same manner as flash tank A.

(a) <u>Wetness y_A leaving the LP throttle</u>

$$y_A = \frac{h_{10} - h_5}{h_{10} - h_9}, \text{ since } h_8 = h_5$$

$$= \frac{88.5}{205.3} = \underline{0.4311}$$

(b) <u>The value of y_B</u>

The energy conservation equation for heat exchangers X_A and X_B together gives:

$$(h_2 - h_3) = (1 - y_B)(h_7 - h_6) + (y_B - y)(h_{11} - h_{10})$$

But $y = y_A y_B = 0.4311\ y_B$

$\therefore\ (y_B - y) = 0.5689\ y_B$

Hence:

$$(380.6 - h_3) = (1 - y_B)(411.5 - 199.0)$$

$$+ 0.5689\, y_B(417.8 - 205.3)$$

$$= 212.5(1 - y_B) + 120.9\, y_B$$

$$h_3 = 168.1 + 91.6\, y_B \qquad (1)$$

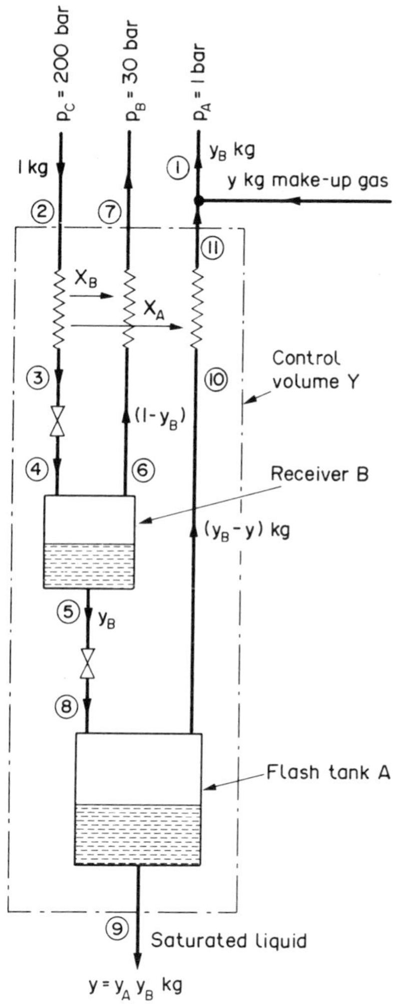

Fig. P.10.10(b)

For the process in receiver (flash tank) B:

$$y_B = \frac{h_6 - h_3}{h_6 - h_5}, \text{ since } h_4 = h_3$$

$$= \frac{199.0 - h_3}{82.2}$$

$$\therefore h_3 = 199.0 - 82.2\, y_B \qquad (2)$$

From eqns. (1) and (2):

$$168.1 + 91.6\, y_B = 199.0 - 82.2\, y_B$$

$$\therefore y_B = \frac{30.9}{173.8} = \underline{0.1778}$$

(c) The yield, y

$$y = 0.4311\, y_B = 0.4311 \times 0.1778 = \underline{0.0766}$$

(d) The value of T_3

From eqn. (2):

$$h_3 = 199.0 - 82.2 \times 0.1778 = 184.4 \text{ kJ/kg}$$

Hence, from Table 18 of Ref. 2, at p = 200 bar, h = 184.4 kJ/kg:

$$T_3 = 160 + \left(\frac{184.4 - 165.5}{187.3 - 165.5}\right) \times 10 = \underline{168.7} \text{ K}$$

(e) Work input per unit yield

For reversible, isothermal compression at T_0 = 290 K:

Compression from state 1 to state 7:

$$\text{Work input per kg} = (b_7 - b_1)$$

$$= T_0(s_1 - s_7) - (h_1 - h_7)$$

$$= 290 \times 0.993 - 6.3 = 281.7 \text{ kJ/kg}$$

$$\therefore [W_{in}]_1^7 = 281.7\, y_B = 281.7 \times 0.1778 = 50.1 \text{ kJ/kg}$$

Compression from state 7 to state 2:

$$[W_{in}]_7^2 = (b_2 - b_7)$$

$$\therefore [W_{in}]_7^2 = T_0(s_7 - s_2) - (h_7 - h_2)$$

$$= 290 \times 0.646 - 30.9 = 156.4 \text{ kJ/kg}$$

$$\therefore \text{Total } W_{in} = 50.1 + 156.4 = 206.5 \text{ kJ/kg delivered}$$

$\therefore$ Work input per unit yield, $w_{in} \equiv W_{in}/y$

$$\therefore w_{in} = \frac{206.5}{0.0766} \times 10^{-3} = \underline{2.696} \text{ MJ/kg}$$

$$= \frac{2.696 \times 10^3}{3600 \times 1.14} = \underline{0.657} \text{ kW h/litre}$$

(f) Rational efficiency

$$\text{Rational efficiency} = \frac{(w_{in})_{REV} \text{ from Problem 10.6}}{w_{in}}$$

$$= \frac{0.701}{2.696} \times 100 = \underline{26.0} \%$$

(i) Heat transfer process in heat exchangers X_A and X_B

Extra work input due to irreversibility $= T_0 \, \Delta S_C$

$$\Delta S_C = [(1 - y_B)(s_7 - s_6) + (y_B - y)(s_{11} - s_{10})] - 1 \times (s_2 - s_3)$$

State 3: $p_3 = 200$ bar, $T_3 = 168.7$ K

From Table 19 of Ref. 2, taking $s \propto \ln T$:

$$s_3 = 1.3223 \text{ kJ/kg K}$$

$$(1 - y_B)(s_7 - s_6) = 0.8222 \times 1.142 = 0.9390 \text{ kJ/kg K}$$

$$(y_B - y)(s_{11} - s_{10}) = 0.1012 \times 1.299 = 0.1315 \text{ kJ/kg K}$$

$$1.0705 \text{ kJ/kg K}$$

$$1 \times (s_2 - s_3) = 2.219 - 1.3223 = 0.8967 \text{ kJ/kg K}$$

$$\therefore \Delta S_C = 0.1738 \text{ kJ/kg K}$$

Per kg of liquid air:

$$\text{Extra work input} = \frac{T_0 \, \Delta S_C}{y} = \frac{290 \times 0.1738}{0.0766} \times 10^{-3} = 0.6580 \text{ MJ/kg}$$

$$\text{Percentage of actual work input} = \frac{0.6580}{2.696} \times 100 = \underline{24.4} \%$$

(ii) HP throttling process

$$\Delta s_C = 1 \times (s_4 - s_3)$$

$$s_4 = s_6 - y_B(s_6 - s_5)$$

$$= 1.723 - 0.1778 \times 0.646 = 1.6081 \text{ kJ/kg K}$$

$$\Delta s_C = 1.6081 - 1.3223 = 0.2858 \text{ kJ/kg K}$$

Per kg of liquid air:

$$\text{Extra work input} = \frac{T_0 \, \Delta s_C}{y} = \frac{290 \times 0.2858}{0.0766} \times 10^{-3} = 1.0820 \text{ MJ/kg}$$

$$\text{Percentage of actual work input} = \frac{1.0820}{2.696} \times 100 = \underline{40.1} \text{ \%}$$

(iii) LP throttling process

$$\Delta s_C = y_B(s_8 - s_5)$$

$$s_8 = s_{10} - y_A(s_{10} - s_9)$$

$$= 2.559 - 0.4311 \times 2.559 = 1.4558 \text{ kJ/kg K}$$

$$\therefore \Delta s_C = 0.1778 \times 0.3788 = 0.06735 \text{ kJ/kg K}$$

Per kg of liquid air:

$$\text{Extra work input} = \frac{T_0 \, \Delta s_C}{y} = \frac{290 \times 0.06735}{0.0766} \times 10^{-3} = 0.2550 \text{ MJ/kg}$$

$$\text{Percentage of actual work input} = \frac{0.2550}{2.696} \times 100 = \underline{9.5} \text{ \%}$$

Check on the rational efficiency

$$\text{Total extra work input} = 24.4 + 40.1 + 9.5 = 74.0 \text{ \%}$$

$$\therefore \text{Rational efficiency} = (100 - 74.0) = \underline{26.0} \text{ \% (as before)}$$

10.11 In a Claude process (Fig. 10.11), the pressure at compressor delivery is 40 bar, the temperature of the air entering the expansion engine is -80 °C, $T_{13} = T_2 = T_1$, $T_9 = T_{10}$ and $x = 0.2$. The compression process is isothermal and reversible, and the isentropic efficiency of the engine process is 75 %.

Calculate (a) the value of T_{10}; (b) the yield y_1; (c) the value of

T_{12}; (d) the wetness of the mixture leaving the throttle; (e) the specific enthalpy of the fluid at points 5 and 4 respectively; (f) the net work input per unit yield, expressed both in MJ/kg and kW h/litre; (g) the rational efficiency of the process.

Solution (Claude process)

State	$\frac{h}{kJ/kg}$	$\frac{s}{kJ/kg\ K}$
1,13	417.8	3.858
2	409.3	2.775
3	299.5	2.312
7	0	0
8	205.3	2.559

(a) The value of T_{10}

For hypothetical isentropic expansion in the expansion engine, the wetness q_{9_s} at exhaust is given by:

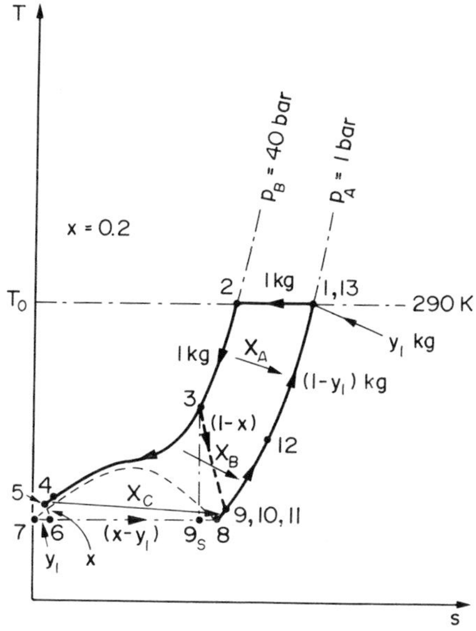

Fig. P.10.11(a)

$$q_{9_s} = \frac{s_8 - s_3}{s_8 - s_7} = \frac{0.247}{2.559} = 0.0965$$

$$h_{9_s} = (1 - q_{9_s})h_8, \text{ since } h_7 = 0$$

$$= 0.9035 \times 205.3 = 185.5 \text{ kJ/kg}$$

$$\Delta h_s \equiv (h_3 - h_{9_s}) = 114.0 \text{ kJ/kg}$$

$$\Delta h = \eta_E \, \Delta h_s = 0.75 \times 114.0 = 85.5 \text{ kJ/kg}$$

$$\therefore h_9 = (h_3 - \Delta h) = 214.0 \text{ kJ/kg}$$

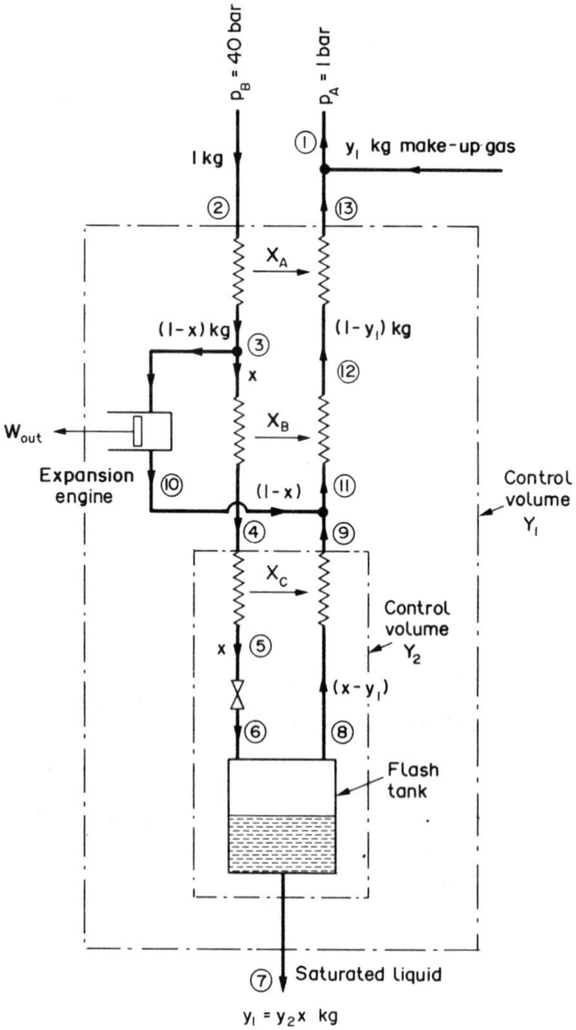

Fig. P.10.11(b)

Now $T_{10} = T_9$ (given). Hence, from Table 18 of Ref. 2 at $p = 1$ bar, $h = 214.0$ kJ/kg:

$$T_{10} = \underline{\mathit{90.0}}\ \text{K}$$

(b) <u>The yield, y_1</u>

The energy conservation equation for control volume Y_1 gives:

$$h_2 = (1 - x)(h_3 - h_{10}) + y_1 h_7 + (1 - y_1)h_{13}$$

$$T_9 = T_{10}\text{(given)} \therefore h_{10} = h_9 = 214.0\ \text{kJ/kg}$$

$$409.3 = 0.8 \times 85.5 + 417.8(1 - y_1)$$

$$(1 - y_1) = \frac{340.9}{417.8} = 0.8159$$

$$\therefore \text{Yield, } y_1 = \underline{\mathit{0.184}}$$

(c) <u>The value of T_{12}</u>

<u>For heat exchanger X_A:</u> $(h_2 - h_3) = (1 - y_1)(h_{13} - h_{12})$

$$109.8 = 0.8159(417.8 - h_{12})$$

$$\therefore h_{12} = 417.8 - 134.6 = 283.2\ \text{kJ/kg}$$

Hence, from Table 18 of Ref. 2, at $p = 1$ bar, $h = 283.2$ kJ/kg:

$$T_{12} = 150 + \left(\frac{283.2 - 276.2}{286.5 - 276.2}\right) \times 10 = \underline{\mathit{156.8}}\ \text{K}$$

(d) <u>The wetness leaving the throttle</u>

Wetness at point 6 $= y_1/x = 0.184/0.2 = \underline{\mathit{0.920}}$

(e) <u>Specific enthalpy at points 5 and 4</u>

<u>For the flash tank:</u>

$$x h_6 = y_1 h_7 + (x - y_1)h_8$$

$$(h_8 - h_6) = \frac{y_1}{x}(h_8 - h_7) = 0.920 \times 205.3 = 188.9\ \text{kJ/kg}$$

$$\therefore h_5 = h_6 = (205.3 - 188.9) = \underline{\mathit{16.4}}\ \text{kJ/kg}$$

For heat exchanger X_C:

$$x(h_4 - h_5) = (x - y_1)(h_9 - h_8)$$

$$(h_4 - 16.4) = \frac{0.016}{0.2} \times (214.0 - 205.3) = 0.7$$

$$\therefore h_4 = \underline{17.1} \text{ kJ/kg}$$

(f) Net work input per unit yield

Since the compression process is stated to be isothermal and reversible:

$$\text{Compressor } W_{in} = T_0(s_1 - s_2) - (h_1 - h_2) = 290 \times 1.083 - 8.5$$

$$= 305.6 \text{ kJ/kg}$$

$$\text{Engine } W_{out} = (1 - x)\Delta h = 0.8 \times 85.5 = 68.4 \text{ kJ/kg}$$

$$\therefore \text{Net } W_{in} = 237.2 \text{ kJ/kg compressed}$$

$$\text{Net work input per unit yield, } w_{in} \equiv \frac{W_{net}}{y_1}$$

$$= \frac{237.2}{0.184} \times 10^{-3} = \underline{1.29} \text{ MJ/kg}$$

$$= \frac{1.29 \times 10^3}{3600 \times 1.14} = \underline{0.314} \text{ kW h/litre}$$

(g) Rational efficiency

From Problem 10.6, the ideal, reversible work input per unit yield is:

$$w_{REV} = 0.701 \text{ MJ/kg}$$

$$\therefore \eta_R \equiv \frac{w_{REV}}{w_{in}} = \frac{0.701}{1.29} \times 100 = \underline{54.3}\ \%$$

10.12 In a Heylandt process (Fig. 10.11), the pressure at compressor delivery is 200 bar, $T_{13} = T_{2'} = T_1$, $T_9 = T_{10}$ and $x = 0.45$. The compression process is isothermal and reversible, and the isentropic efficiency of the expansion engine is 75 %.

Calculate (a) the value of T_{10}; (b) the yield y_1; (c) the wetness of the mixture leaving the throttle; (d) the values of T_5 and T_4 respectively; (e) the net work input per unit yield, expressed both in MJ/kg and kW h/litre; (f) the rational efficiency of the process.

Evaluate the extra work input due to irreversibility arising from (i) the heat transfer processes in heat exchangers X_B and X_C respectively, (ii) the throttling process, (iii) the engine process, expressing these quantities as a percentage of the actual work input.

Solution (Heylandt process)

State	$\frac{h}{kJ/kg}$	$\frac{s}{kJ/kg\ K}$
1,12,13	417.8	3.858
2',3	380.6	2.219
7	0	0
8	205.3	2.559

(a) The value of T_{10}

For hypothetical isentropic expansion in the expansion engine, the wetness q_{9_s} at exhaust is given by:

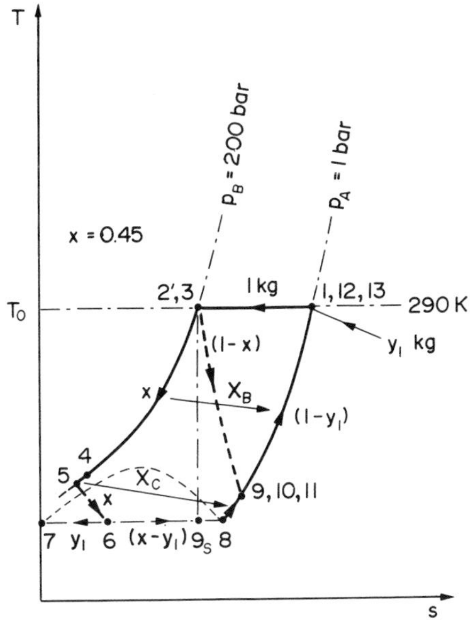

Fig. P.10.12(a)

$$q_{9_s} = \frac{2.559 - 2.219}{2.559} = 0.1329$$

$$h_{9_s} = (1 - q_{9_s})h_8, \text{ since } h_7 = 0$$

$$= 0.8671 \times 205.3 = 178.0 \text{ kJ/kg}$$

$$\Delta h_s \equiv (h_3 - h_{9_s}) = 202.6 \text{ kJ/kg}$$

$$\Delta h = \eta_E \, \Delta h_s = 0.75 \times 202.6 = 152.0 \text{ kJ/kg}$$

$$\therefore \; h_9 = (h_3 - \Delta h) = 228.6 \text{ kJ/kg}$$

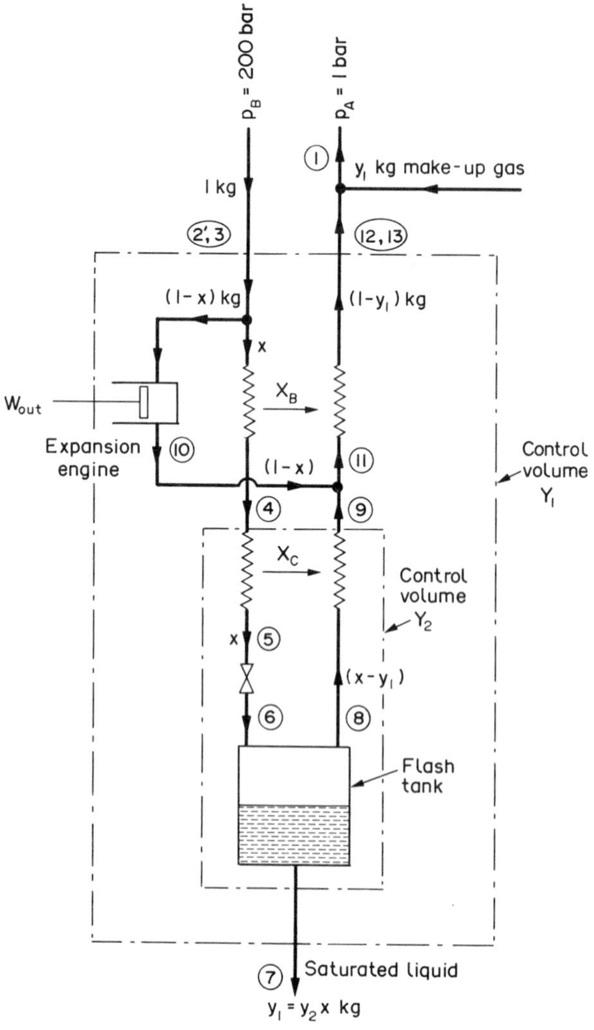

Fig. P.10.12(b)

Now $T_{10} = T_9$. Hence, from Table 18 of Ref. 2, at p = 1 bar, h = 228.6 kJ/kg:

$$T_{10} = 100 + \frac{4.2}{10.4} \times 10 = \underline{104.0} \text{ K}$$

(b) The yield, y_1

The energy conservation equation for control volume Y_1 gives:

$$h_3 = (1 - x)(h_3 - h_{10}) + y_1 h_7 + (1 - y_1)h_{12}$$

$$T_9 = T_{10} \text{ (given)} \therefore h_{10} = h_9 = 228.6 \text{ kJ/kg}$$

$$380.6 = 0.55 \times 152.0 + 417.8(1 - y_1)$$

$$(1 - y_1) = \frac{297.0}{417.8} = 0.7109$$

$$\text{Yield, } y_1 = \underline{0.289}$$

(c) The wetness leaving the throttle

$$\text{Wetness at point 6} \equiv y_2 = y_1/x = 0.2891/0.45 = \underline{0.6424}$$

(d) The values of T_5 and T_4

For the flash tank:

$$(h_8 - h_6) = \frac{y_1}{x}(h_8 - h_7) = \frac{0.2891}{0.45} \times 205.3 = 131.9 \text{ kJ/kg}$$

$$\therefore h_5 = h_6 = 205.3 - 131.9 = 73.4 \text{ kJ/kg}$$

Hence, from Table 18 of Ref. 2, at p = 200 bar, $h_5 = 73.4$ kJ/kg:

$$T_5 = 120 + \frac{1.3}{11.9} \times 5 = \underline{120.5} \text{ K}$$

For heat exchanger X_C:

$$x(h_4 - h_5) = (x - y_1)(h_9 - h_8)$$

$$0.45(h_4 - 73.4) = 0.1609(228.6 - 205.3)$$

$$h_4 - 73.4 = 8.3 \quad \therefore h_4 = 81.7 \text{ kJ/kg}$$

Hence, from Table 18 of Ref. 2, at p = 200 bar, h = 81.7 kJ/kg:

$$T_4 = 120 + \frac{9.6}{11.9} \times 5 = \underline{124.0} \text{ K}$$

(e) Net work input per unit yield

For reversible, isothermal compression, from Problem 10.8:

$$\text{Compressor } W_{in} = T_0(s_1 - s_{2'}) - (h_1 - h_{2'}) = 438.1 \text{ kJ/kg}$$

$$\text{Engine } W_{out} = (1 - x)\Delta h = 0.55 \times 152.0 = 83.6 \text{ kJ/kg}$$

$$\therefore \text{ Net } W_{in} = 354.5 \text{ kJ/kg compress}$$

Net work input per unit yield, $w_{in} = \frac{W_{net}}{y_1}$

$$= \frac{354.5}{0.2891} \times 10^{-3} = \underline{1.226} \text{ MJ/kg}$$

$$= \frac{1.226 \times 10^3}{3600 \times 1.14} = \underline{0.299} \text{ kW h/litre}$$

(f) Rational efficiency

From Problem 10.6, the ideal, reversible work input per unit yield is:

$$w_{REV} = 0.701 \text{ MJ/kg}$$

$$\therefore \eta_R \equiv \frac{w_{REV}}{w_{in}} = \frac{0.701}{1.226} \times 100 = \underline{57.2} \text{ \%}$$

(i) Heat transfer processes in heat exchangers X_B and X_C

From Table 19 of Ref. 2, and noting that s α ln T is a better approximation than an assumption that s α T:

State 4: $p_4 = 200$ bar, $T_4 = 124.0$ K, $s_4 = 0.611$ kJ/kg K

State 5: $p_5 = 200$ bar, $T_5 = 120.5$ K, $s_5 = 0.542$ kJ/kg K

State 11: $p_{11} = 1$ bar, $T_{11} = T_{10} = 104.0$ K, $s_{11} = 2.810$ kJ/kg

Extra work input due to irreversibility = $T_0 \; \Delta S_C$,

where ΔS_C is the entropy creation in the process.

Heat exchanger X_B

$$\Delta s_C = (1 - y_1)(s_{12} - s_{11}) - x(s_3 - s_4)$$

$$(1 - y_1)(s_{12} - s_{11}) = 0.7109(3.858 - 2.810) = 0.7450 \text{ kJ/kg K}$$

$$x(s_3 - s_4) = 0.45(2.219 - 0.611) = 0.7236 \text{ kJ/kg K}$$

$$\therefore \Delta s_C = 0.0214 \text{ kJ/kg K}$$

Per kg of liquid air:

$$\text{Extra work input} = \frac{T_0 \Delta s_C}{y_1} = \frac{290 \times 0.0214}{0.2891} \times 10^{-3} = 0.0215 \text{ MJ/kg}$$

$$\text{Percentage of actual work input} = \frac{0.0215}{1.226} \times 100 = \underline{1.75}\ \%$$

Heat exchanger X_C

$$\Delta s_C = (x - y_1)(s_9 - s_8) - x(s_4 - s_5), \quad \text{and } s_9 = s_{10} = s_{11}$$

$$(x - y_1)(s_9 - s_8) = 0.1609(2.810 - 2.559) = 0.0404 \text{ kJ/kg K}$$

$$x(s_4 - s_5) = 0.45(0.611 - 0.542) = 0.0311 \text{ kJ/kg K}$$

$$\therefore \Delta s_C = 0.0093 \text{ kJ/kg K}$$

Per kg of liquid air:

$$\text{Extra work input} = \frac{T_0 \Delta s_C}{y_1} = \frac{290 \times 0.0093}{0.2891} \times 10^{-3} = 0.0093 \text{ MJ/kg}$$

$$\text{Percentage of actual work input} = \frac{0.0093}{1.226} \times 100 = \underline{0.75}\ \%$$

(ii) The throttling process

$$s_6 = (1 - 0.6424) \times 2.559 = 0.915 \text{ kJ/kg K}$$

$$s_5 = 0.542 \text{ kJ/kg K}$$

$$(s_6 - s_5) = 0.373 \text{ kJ/kg K}$$

$$\therefore \Delta s_C = x(s_6 - s_5) = 0.45 \times 0.373 = 0.1679 \text{ kJ/kg K}$$

Per kg of liquid air:

$$\text{Extra work input} = \frac{T_0 \Delta s_C}{y_1} = \frac{290 \times 0.1679}{0.2891} \times 10^{-3} = 0.1684 \text{ MJ/kg}$$

$$\text{Percentage of actual work input} = \frac{0.1684}{1.226} \times 100 = \underline{13.7}\ \%$$

(iii) The engine process

$$\Delta S_C = (1 - x)(s_{10} - s_3) = 0.55(2.810 - 2.219) = 0.3251 \text{ kJ/}$$

Per kg of liquid air:

$$\text{Extra work input} = \frac{T_0\ \Delta S_C}{y_1} = \frac{290 \times 0.3251}{0.2891} \times 10^{-3} = 0.3261 \text{ MJ/kg}$$

$$\text{Percentage of actual work input} = \frac{0.3261}{1.226} \times 100 = \underline{26.6}\ \%$$

Check on the rational efficiency

$$\text{Total extra work input} = 1.75 + 0.75 + 13.7 + 26.6 = 42.8\ \%$$

$$\therefore \text{ Rational efficiency, } \eta_R = (100 - 42.8) = \underline{57.2}\ \% \text{ (as before)}$$

10.13 Draw pressure-volume and temperature-entropy diagrams for the reversed Stirling cycle and the reversed Ericsson cycle (Section 10.19).

Solution

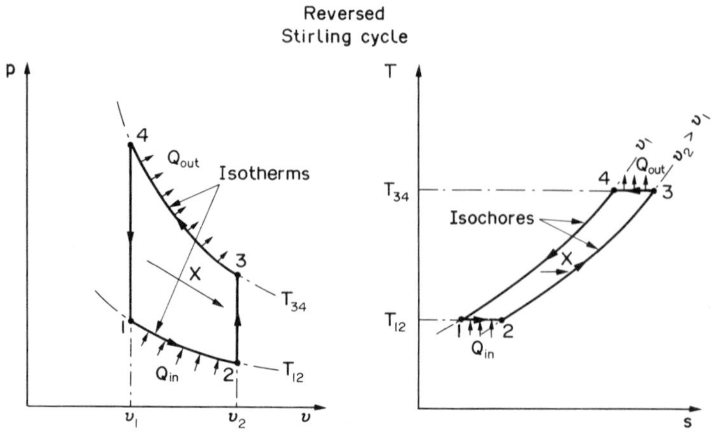

Fig. P.10.13(a)

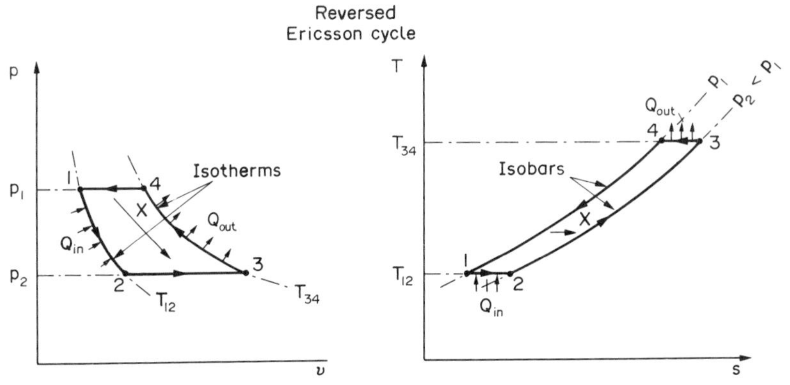

Fig. P.10.13(b)

REFERENCES

1. Haywood, R. W., *Analysis of Engineering Cycles*, Pergamon Press, Oxford, 3rd ed., 1980/1985. (2nd ed. also in Russian translation - R.V.KHEJVUD, *Analiz tsiklov v tekhnicheskoj termodinamike*. Perevod s anglijskogo E.Ya.Gadasa, Moskva, "Energiya", 1979. 3rd. ed. also in Spanish translation - R. W. Haywood, *Analisis Termodinamico de Plantas Electricas*. Translated by Ing. Mario Sanchez Orozco, Compania Editorial Limusa, S.A., México, 1986.)

2. Haywood, R. W., *Thermodynamic Tables in SI (metric) Units*, with enthalpy-entropy diagram for steam and pressure-enthalpy diagram for Refrigerant-12, Cambridge University Press, 2nd ed., 1972/1981. [Also in Spanish translation - *Tablas de Termodinámica en Unidades SI (métricas)*. Translated by A.E. Estrada, Compania Editorial Continental, S.A., México, 1977.]

3. Haywood, R. W., *Equilibrium Thermodynamics for Engineers and Scientists*, John Wiley & Sons Ltd., Chichester, 1980 (Also in Russian translation - R. V. KHEIVUD, *Termodinamika ravnovesnykh protsessov*. Perevod s anglijskogo V. F. Pastushenko, Moskva, "Mir", 1983.)

4. *UK Steam Tables in SI Units 1970*, Ed. Arnold (Publishers) Ltd., London, 1970.